AF388671

Food Wastes

Food Wastes

Feedstock for Value-Added Products

Special Issue Editor

Diomi Mamma

MDPI • Basel • Beijing • Wuhan • Barcelona • Belgrade • Manchester • Tokyo • Cluj • Tianjin

Special Issue Editor
Diomi Mamma
National Technical University of Athens
Greece

Editorial Office
MDPI
St. Alban-Anlage 66
4052 Basel, Switzerland

This is a reprint of articles from the Special Issue published online in the open access journal *Fermentation* (ISSN 2311-5637) (available at: https://www.mdpi.com/journal/fermentation/special issues/ferment food wastes).

For citation purposes, cite each article independently as indicated on the article page online and as indicated below:

LastName, A.A.; LastName, B.B.; LastName, C.C. Article Title. *Journal Name* **Year**, *Article Number*, Page Range.

ISBN 978-3-03936-419-0 (Hbk)
ISBN 978-3-03936-420-6 (PDF)

Contents

About the Special Issue Editor

Diomi Mamma, Ph.D., is an Assistant Professor of Bioprocess Engineering at the School of Chemical Engineering, National Technical University of Athens, Greece. She studied Chemical Engineering at the School of Chemical Engineering, National Technical University of Athens, where she obtained her Ph.D. (2002). She teaches subjects related to bioprocess engineering. Her research interests focus on microbial production and characterization of enzymes, bioconversion of biomass to ethanol applying different fermentation strategies, and environmental biotechnology, with emphasis on the design of appropriate biological processes for the complete removal of xenobiotics.

 *fermentation*

Editorial

Food Wastes: Feedstock for Value-Added Products

Diomi Mamma

Biotechnology Laboratory, School of Chemical Engineering, National Technical University of Athens Zografou Campus, 15780 Athens, Greece; dmamma@chemeng.ntua.gr

Received: 22 April 2020; Accepted: 24 April 2020; Published: 27 April 2020

Food is a precious commodity, and its production can be resource-intensive. According to the Food and Agriculture Organization of the United Nations, nearly 1.3 billion tons of food products per year are lost along the food supply chain, and in the next 25 years the amount of food waste has been projected to increase exponentially. Food waste is produced at any stage of the supply chain, which extends from the agricultural site to the processing plant and finally the retail market. The management of food waste should follow certain policies based on the 3R's concept, i.e., reduce, reuse, and recycle [1]. Generally, food waste is composed of a heterogeneous mixture formed by carbohydrates (starch, cellulose, hemicellulose, or lignin), proteins, lipids, organic acids, and smaller inorganic parts. Currently, most food wastes are recycled, mainly as animal feed and compost. The remaining quantities are incinerated and disposed in landfills, causing serious emissions of methane (CH_4), which is 23 times more potent than carbon dioxide (CO_2) as a greenhouse gas and significantly contributes to climate change [2]. Valorizing food waste components could in fact lead to numerous possibilities for the production of valuable chemicals, fuels, and products [1].

The present Special Issue compiles a wide spectrum of aspects of research and technology in the area of "food waste exploitation", and highlights prominent current research directions in the field for the production of value-added products such as polylactic acid, hydrogen, ethanol, enzymes, and edible insects.

Polylactic acid (PLA) is a biodegradable polymer with great potential in replacing petrochemical polymers. The morphological, mechanical, and thermal properties of the polymer are determined by the presence of different amounts of L- and D-lactic acid monomers or oligomers [3]. The microbial production of optically pure lactic acid has extensively been studied, because chemically synthesized lactic acid is a racemic mixture. Optimizing culture conditions and selecting the LAB strains capable of producing D-lactic acid with high yield and optical purity from orange peel waste as raw material can contribute to the development of biowaste refineries. Bustamante et al. [4] evaluated six strains of the species *Lactobacillus delbrueckii* ssp. *bulgaricus* for the production of D-lactic acid from orange peel waste hydrolysate. *L. delbrueckii ssp. bulgaricus* CECT 5037 had the best performance, with a yield of 84% w/w for D-LA production and up to 95% enantiomeric excess (optical purity).

Biomethanation (methane fermentation) is a complex biological process, which can be divided in four phases of biomass degradation and conversion, namely, hydrolysis, acidogenesis, acetogenesis, and methanation. The individual phases are carried out by different groups of micro-organisms (bacteria), which partly stand in syntrophic interrelation and place different requirements on the environment. Undissolved compounds like cellulose, proteins, and fats are hydrolyzed into monomers by enzymes produced by facultative and obligatorily anaerobic bacteria [5]. The use of a microbial consortium consisting of the microbial flora of methane production and microorganisms that can degrade cellulosic biomass like *Clostridium cellulovorans* was proven efficient in degraded mandarin orange peel without any pretreatments and produced methane that accounted for 66.2% of the total produced gas [6].

Hydrogen is a noncarbonaceous fuel and energy carrier possessing higher net calorific value compared to other fuels (120 MJ/kg versus 46.7 MJ/kg for gasoline). Microbes primarily produce hydrogen via photofermentation by the purple nonsulfur bacteria *Rhodobacter* and *Rhodopseudomonas*, and during dark fermentation by strictly anaerobic *Clostridium* species [7,8]. Depending upon the availability of substrate, the selection of functional microorganisms necessary for hydrogen production is an important step. Simulation of the exchange metabolic fluxes of monocultures and pairwise cocultures using genome-scale metabolic models on artificial garbage slurry resulted in the identification of one of the top hydrogen producing cocultures comprising *Clostridium beijerinckii* NCIMB 8052 and *Yokenella regensburgei* ATCC 43003. The consortium produced a similar amount of hydrogen gas and increased butyrate (attributed to cross-feeding of lactate produced by *Y. regensburgei*), compared to the *C. beijerinckii* monoculture, when grown on the artificial garbage slurry [9].

Household food waste is a complex biomass containing various components that make it a source of potential fermentative substrates. The general scheme of bioethanol production from such complex materials involves a pretreatment step that increases the digestibility of the material—enzymatic hydrolysis—to liberate the monosaccharides and fermentation of these sugars to ethanol. In terms of cost, the most demanding step, which significantly increases the total cost of the production of bioethanol and is identified as a barrier in the further deployment of ethanol production, is enzymatic hydrolysis. If the necessary enzymes could be efficiently produced on-site, the cost could be significantly reduced. A recent study has estimated that the cellulase cost can be reduced from 0.78 to 0.58$/gallon by shifting from the off-site to the on-site approach of cellulase production [10]. The mesophilic fungus *Fusarium oxysporum* F3 grown under solid state cultivation on wheat bran produced a multienzyme system capable of hydrolyzing the carbohydrates present in household food waste. The use of mixed-microbial cultures in bioethanol production step consisting of *F. oxysporum* solid state culture and the yeast *Saccharomyces cerevisiae* increased bioethanol volumetric productivity, compared to mono-culture of the fungus. Bioethanol production increased by approximately 23% when the mixed microbial culture was supplemented with low dosages of commercial glucoamylase [11].

Carrión-Paladines et al. [12] evaluated two *Xylaria* spp. of the dry forest areas of southern Ecuador, for ligninase and cellulase production under solid state fermentation using residues obtained from the Palo Santo essential oil extraction. The Palo Santo is considered a vital resource for the local communities of the dry forest, as different parts of the tree are used in traditional medicine, as well as for the extraction of essential oil. The essential oil extraction process generates abundant organic waste, which is commonly discarded directly into the natural ecosystems or burned. Laccase, cellulose, and xylanase activities of *Xylaria feejeensis* and *Xylaria* cf. *microceras* were generally higher than those of the control fungus *Trametes versicolor* (L.) Lloyd, furthering the understanding of the potential use of native fungi as ecologic lignocellulosic decomposers and for industrial proposes.

Beer production generates large quantities of spent yeast during the fermentation and lagering process. The spent yeast is an efficient starting material to produce yeast extract, which is generally defined as the soluble content of a yeast cell that remains once the cell wall has been destroyed and removed. The variety of different physiologically valuable substances in yeast cells offer the possibility of use as a yeast extract in different areas of the food industry. Jacob et al. [13] demonstrated that the composition of various physiologically valuable substance groups of a yeast extract depends on the biodiversity of the spent yeast from beer production, indicating that brewer's spent yeast should be carefully selected to produce a yeast extract with a defined nutritional composition.

In many cases, food wastes are difficult to utilize for the recovery of value-added products due to their biological instability or potentially pathogenic nature. Fusarium head blight (FHB), a fungal disease caused by several *Fusarium* spp., is one of the most significant causes of economic loss in cereal crops. *Fusarium* spp. produce various amounts and types of trichothecene mycotoxins, with deoxynivalenol being the major one, which are highly toxic to humans and livestock. A method to recover the nutrients from the affected cereals, without the mycotoxins, was reported by Gulsunoglu et al. [14]. The infected grains were initially fermented under solid state cultivation with *Aspergillus oryzae* and/or

Lactobacillus plantarum. The fermented material was provided to black soldier fly larvae, which consumed deoxynivalenol-contaminated materials and converted them in insect biomass without accumulating deoxynivalenol in their bodies. This treatment technology using black soldier fly larvae may contribute to reducing the burden of animal protein shortages in the animal feed market.

Varelas [15] compiled up-to-date information on the mass rearing of edible insects for food and feed based on food wastes. Edible insects are insect species that can be used for human consumption but also for livestock feed as a whole, parts of them, and/or protein, and lipid extract.

Funding: This research received no external funding.

Acknowledgments: The editor wish to thank our article contributors, Editorial Board members, Reviewers, and Assistant Editors of this journal, whose contributions made the publication of this Special Issue possible.

Conflicts of Interest: The authors declare no conflict of interest.

References

1. Nayak, A.; Bhushan, B. An overview of the recent trends on the waste valorization techniques for food wastes. *J. Environ. Manag.* **2019**, *233*, 352–370. [CrossRef] [PubMed]
2. Lin, C.S.K.; Pfaltzgraff, L.A.; Herrero-Davila, L.; Mubofu, E.B.; Abderrahim, S.; Clark, J.H.; Koutinas, A.A.; Kopsahelis, N.; Stamatelatou, K.; Dickson, F.; et al. Food waste as a valuable resource for the production of chemicals, materials and fuels. Current situation and global perspective. *Energy Environ. Sci.* **2013**, *6*, 426–464. [CrossRef]
3. Singhvi, M.; Zendo, T.; Sonomoto, K. Free lactic acid production under acidic conditions by lactic acid bacteria strains: Challenges and future prospects. *Appl. Microbiol. Biotechnol.* **2018**, *102*, 1–14. [CrossRef] [PubMed]
4. Bustamante, D.; Tortajada, M.; Ramón, D.; Rojas, A. Production of D-Lactic Acid by the Fermentation of Orange Peel Waste Hydrolysate by Lactic Acid Bacteria. *Fermentation* **2020**, *6*, 1. [CrossRef]
5. Xu, N.; Liu, S.; Xin, F.; Zhou, J.; Jia, H.; Xu, J.; Jiang, M.; Dong, W. Biomethane Production from Lignocellulose: Biomass Recalcitrance and Its Impacts on Anaerobic Digestion. *Front. Bioeng. Biotechnol.* **2019**, *7*, 191. [CrossRef] [PubMed]
6. Tomita, H.; Tamaru, Y. The Second-Generation Biomethane from Mandarin Orange Peel under Cocultivation with Methanogens and the Armed *Clostridium Cellulovorans*. *Fermentation* **2019**, *5*, 95. [CrossRef]
7. Łukajtis, R.; Hołowacz, I.; Kucharska, K.; Glinka, M.; Rybarczyk, P.; Przyjazny, A.; Kamińskia, M. Hydrogen production from biomass using dark fermentation. *Renew. Sustain. Energy Rev.* **2018**, *91*, 665–694. [CrossRef]
8. Keskin, T.; Abo-Hashesh, M.; Hallenbeck, P.C. Photofermentative hydrogen production from wastes. *Bioresour. Technol.* **2011**, *102*, 8557–8568. [CrossRef] [PubMed]
9. Schwalm, N.D., III; Mojadedi, W.; Gerlach, E.S.; Benyamin, M.; Perisin, M.A.; Akingbade, K.L. Developing a Microbial Consortium for Enhanced Metabolite Production from Simulated Food Waste. *Fermentation* **2019**, *5*, 98. [CrossRef]
10. Johnson, E. Integrated enzyme production lowers the cost of cellulosic ethanol. *Biofuels Bioprod. Biorefin.* **2016**, *10*, 164–174. [CrossRef]
11. Prasoulas, G.; Gentikis, A.; Konti, A.; Kalantzi, S.; Kekos, D.; Mamma, D. Bioethanol Production from Food Waste Applying the Multienzyme System Produced On-Site by *Fusarium oxysporum* F3 and Mixed Microbial Cultures. *Fermentation* **2020**, *6*, 39. [CrossRef]
12. Carrión-Paladines, V.; Fries, A.; Caballero, R.E.; Pérez Daniëls, P.; García-Ruiz, R. Biodegradation of Residues from the Palo Santo (Bursera graveolens) Essential Oil Extraction and Their Potential for Enzyme Production Using Native Xylaria Fungi from Southern Ecuador. *Fermentation* **2019**, *5*, 76. [CrossRef]
13. Jacob, F.F.; Striegel, L.; Rychlik, M.; Hutzler, M.; Methner, F.-J. Spent Yeast from Brewing Processes: A Biodiverse Starting Material for Yeast Extract Production. *Fermentation* **2019**, *5*, 51. [CrossRef]

14. Gulsunoglu, Z.; Aravind, S.; Bai, Y.; Wang, L.; Kutcher, H.R.; Tanaka, T. Deoxynivalenol (DON) Accumulation and Nutrient Recovery in Black Soldier Fly Larvae (*Hermetia illucens*) Fed Wheat Infected with *Fusarium* spp. *Fermentation* **2019**, *5*, 83. [CrossRef]

15. Varelas, V. Food Wastes as a Potential New Source for Edible Insect Mass Production for Food and Feed: A review. *Fermentation* **2019**, *5*, 81. [CrossRef]

 *fermentation*

Article

Production of D-Lactic Acid by the Fermentation of Orange Peel Waste Hydrolysate by Lactic Acid Bacteria

Daniel Bustamante [1,2], Marta Tortajada [2], Daniel Ramón [2] and Antonia Rojas [2,*]

1   Current address: National Renewable Energy Centre (CENER), Av. Ciudad de la Innovación, 7, 31621 Sarriguren, Spain; dbustamante@cener.com
2   ADM-BIOPOLIS, Parc Científic Universitat de València, C/Catedrático Agustín Escardino, 9, 46980 Paterna, Spain; marta.tortajada@adm.com (M.T.); daniel.ramonvidal@adm.com (D.R.)
*   Correspondence: antonia.rojas@adm.com

Received: 31 October 2019; Accepted: 16 December 2019; Published: 18 December 2019

Abstract: Lactic acid is one the most interesting monomer candidates to replace some petroleum-based monomers. The application of conventional poly-lactic acid (PLA) is limited due to insufficient thermal properties. This limitation can be overcome by blending poly-D and poly-L-lactic acid. The main problem is the limited knowledge of D-lactic acid (D-LA) production. Efficient biochemical processes are being developed in order to synthesize D-LA from orange peel waste (OPW). OPW is an interesting renewable raw material for biorefinery processes of biocatalytic, catalytic or thermal nature owing to its low lignin and ash content. Bioprocessing of the pretreated OPW is carried out by enzymatic hydrolysis and fermentation of the released sugars to produce D-LA. Several strains of the species *Lactobacillus delbrueckii* ssp. *bulgaricus* have been evaluated for the production of D-LA from OPW hydrolysate using *Lactobacillus delbrueckii* ssp. *delbrueckii* CECT 286 as a reference strain since its performance in this kind of substrate have been widely reported in previous studies. Preliminary results show that *Lactobacillus delbrueckii* ssp. *bulgaricus* CECT 5037 had the best performance with a yield of 84% *w/w* for D-LA production and up to 95% (e.e.).

Keywords: added value product; D-lactic acid; LAB strains; food waste; orange peel waste

---

1. Introduction

Lactic acid is an important chemical and has attracted a great attention due its widespread applications in the food, pharmaceutical, cosmetic, and textile industries. Polylactic acid (PLA) is a biodegradable polymer with great potential in replacing petrochemical polymers and therefore, L-and D-lactic acids are prominent monomers of the bioplastic industry [1]. The morphological, mechanical and thermal properties of the polymer are determined by the presence of different amounts of L- and D-lactic acid monomers or oligomers [2–6]. Microbial production of optically pure lactic acid has extensively been studied because chemically synthesized lactic acid is a racemic mixture [7]. In fact, the optimization of operation conditions is very effective to achieve high selectivity to the isomer of interest [8]. Although the L-isomer has been studied in detail, information on biosynthesis of D-lactic acid (D-LA) is still limited [5,9].

PLA market demand accounts for 11.4% of total bioplastic production worldwide, approximately 18×10^4 metric tons per year and the PLA demand is estimated to grow by 28% per year until 2025. However, production costs of PLA are still high, mainly due to expensive fermentation media components. To overcome this problem, several residues have been employed as raw material [3,5,7,10–12]. Production of D-LA from liquid pineapple wastes [13], date juice [14], corn stover [15], hardwood pulp hydrolysate [16] and brown rice [17] has been studied. In this sense, the valorization of food waste to useful products such as D-LA is a good alternative [1,18,19]. In particular, orange peel and pulp waste (OPW) can be used to produce D-LA after adequate pre-treatment processes [20–22].

Orange waste is the most abundant citrus waste with up to 50 million metric tons of oranges consumed every year [23]. This huge amount of waste accounts for 45%–60% of the total fruit weight, and therefore, a lot of potential applications have been studied for their valorization to date [24]. The main application of this residue is as an ingredient for cattle feed or as pelletized dry solid fuel, but its processing results in highly polluted wastewater [25]. The use of citrus waste to produce compounds of high added value, essential oils, fertilizer, pectin, industrial enzymes, ethanol and absorbents has recently been described [21,23–28]. In addition, orange waste present low levels of lignin and a large amount of sugars [27], which make it an ideal substrate for fermentation processes after the implementation of the required pre-treatment and enzymatic hydrolysis stages.

Lactic acid is produced in high amounts by lactic acid bacteria (LAB) which can do so in a homofermentative way employing the Embden-Meyerhof pathway where lactic acid is the only acid produced, or by the heterofermentative way following the phosphogluconate and phosphoketolase pathway where lactic acid is one of the products and yields of 0.5 g g^{-1} of hexose. LABs produce either one or the two forms of lactate [4,11,29,30]. The species *Lactobacillus delbrueckii* ssp. *delbrueckii* has been reported as a homofermentative producer of D-LA using several agro-industrial residues [9]. This bacterium yields 90% D-LA from sugarcane molasses, 95% D-LA from sugarcane juice, 88% D-LA from sugar beet juice [31] and 88% D-LA from orange peel waste (OPW) [32]. Moreover, the species *Lactobacillus delbrueckii* subsp. *bulgaricus* has been used in the dairy industry to transform milk into yogurt and some strains are able to produce highly pure D-LA [33]. Therefore, lactose and whey have been widely studied as raw materials for lactic acid production [34–36], even cloning the D-lactate dehydrogenase gene in *Escherichia coli* [37]. Other studies included wheat flour, molasses, sorghum and lignocellulosic hydrolysates as feedstocks for the production of lactic acid by *Lactobacillus delbrueckii* subsp. *bulgaricus*, especially for L-LA isomer production [11,38]. This fact means that some strains of *Lactobacillus delbrueckii* subsp. *bulgaricus* could be potential candidates for D-LA production from sustainable feedstocks.

The aim of this work was to find LAB strains capable of producing D-LA with high yield and optical purity from OPW as raw material to contribute in the development of biowaste-refineries. For this purpose, several *Lactobacillus delbrueckii* ssp. *bulgaricus* strains were evaluated in comparison to the reference strain *Lactobacillus delbrueckii* ssp. *delbrueckii* CECT 286 which has been reported as a high yield producer of D-LA from biowaste and OPW hydrolysate in particular.

2. Materials and Methods

2.1. Bacterial Strains, Media and Growth Conditions

The bacterial strains employed in this study are listed in Table 1 and *Lactobacillus delbrueckii* ssp. *delbrueckii* CECT 286 was used as reference strain. The selected strains were purchased from the Spanish Type Culture Collection (CECT). After being received they were recovered in MRS medium and stored in 20% glycerol at −80 °C for long-term preservation. Precultures were prepared in tubes containing MRS medium with a small headspace and incubated overnight at 37 °C and static micro-aerobic conditions.

Table 1. Lactic acid bateria (LAB) strains selected for D-lactic acid production screening.

Microorganism	Strain Code
L. delbrueckii ssp. *bulgaricus*	CECT 4005
L. delbrueckii ssp. *bulgaricus*	CECT 4006
L. delbrueckii ssp. *bulgaricus*	CECT 5035
L. delbrueckii ssp. *bulgaricus*	CECT 5036
L. delbrueckii ssp. *bulgaricus*	CECT 5037
L. delbrueckii ssp. *bulgaricus*	CECT 5038

Screening of LAB strains was performed in 15 mL tubes at 37 °C and using a medium with sugars resembling OPW hydrolysate as follows: MRS broth plus glucose 30 g L^{-1}, fructose 20 g L^{-1}, galactose 5 g L^{-1} and arabinose 6 g L^{-1}. Cultures were inoculated in duplicate with 5% *v/v* of preculture and were incubated in orbital shaker at 200 rpm. Aerobic and micro-aerobic conditions were tested at pH 6.2 for 40 h.

2.2. OPW Hyrolysate Tolerance Assays

Tolerance assays were performed in triplicate using selected strains and preparing a multi-well plate with 200 μL of MRS with OPW hydrolysate diluted at 50%, 85% and 100% *v/v* as culture medium for each condition. Precultures were prepared in MRS and inoculated at 10% of total volume. A microplate incubator spectrophotometer was used with temperature set at 37 °C for 45 h. The plate was shaken every hour for 5 seconds before each OD$_{600}$ measurement to obtain the growth curves of the strains.

2.3. Fermentation Assays

Strains were cultured in 50 mL tubes containing MRS with 85% *v/v* OPW hydrolysate at pH 6.2, 37 °C and 45 °C in micro-aerobic conditions. An additional assay was done by adjusting pH at 5.8 each 24 h with NaOH 5 M. All runs started by inoculating 15% *v/v* of preculture and then incubated in an orbital shaker at 200 rpm for 120 h.

The experiments in the bioreactor setup were performed in 1.5 L Applikon$^{®}$ in batch mode with OPW hydrolysate at 85% *v/v* with MRS and 5 g L^{-1} meat extract as additional nitrogen source. The OPW hydrolysate was sterilized using sterile glass fiber and cellulose acetate membrane filters with 0.2 μm of pore size, and then added to the bioreactor. Before the inoculum addition, the anaerobic atmosphere was obtained by stripping the oxygen off with a nitrogen stream. The experimental conditions were set up at 37 °C, 200 rpm, and pH of 5.8, adding NaOH 5 M or HCl 2 M for pH control during fermentation.

2.4. OPW Pretreatments

The substrate used in this study was OPW obtained from juice elaboration. These residues were blade-milled to a final particle diameter of around 5 mm and then, samples were subsequently stored in a freezer at −20 °C until use. The characterization of the raw material was performed according to the NREL procedures for determination of structural carbohydrates and free sugars, in addition to extractives [39–41], while moisture was assessed by using an infrared drying balance at temperatures between 70 and 90 °C until constant weight. The results obtained by applying the NREL methodology are compiled in Table 2. For D-LA production assays, OPW was milled down to 1–2 mm particle size and hydrolysis was carried out at 10% *w/w* of dry solid, 50 °C, 300 rpm and initial pH of 5.2 using enzyme cocktails with cellulases, β-glucosidase, xylanase, β-xylosidase, pectinase, and auxiliary activities (Celluclast 1.5 l, Novozym 188, Pectinex Ultra SP-L gifted by Novozymes) as described by de la Torre and colleagues [22].

Table 2. OPW composition analysis according to NREL protocols.

Component	% Dry Weight (*w/w*)
Total solids	19.2 ± 0.5
Ash	3.9 ± 0.2
Fats	n.d.
Water extractives	37.5 ± 0.4
Free sugars	36.4 ± 0.6
Glucan	19.1 ± 0.1
Hemicellulose	14.8 ± 0.2
Lignin	6.2± 0.5
Pectin	17.9 ± 1.5

2.5. Analytical Procedures

The content of sugars and organic acids was determined by HPLC liquid chromatography (2695 HPLC with a refractive Index Detector 2414; Waters, Cerdanyola del Vallés, Spain) using a Rezex ROA Organic acid column, with H_2SO_4 at 2.5 mM and 0.5 mL min^{-1} flow. The optical purity of D-LA was determined by HPLC (Agilent Technologies 1100 Series, Waldbronn, Germany) using a DAD detector, a Chirex 3126 (D)-penicillamine (250 × 4.6; Phenomenex) column working at room temperature, and a $CuSO_4$ 1 mM solution as mobile phase flowing at 1.2 mL min^{-1}.

3. Results and Discussion

3.1. Screening of LAB Strains for D-LA Production

Lactic acid production was tested in 15 mL tubes containing 3 mL of culture resembling OPW hydrolysate for aerobic conditions and 14 mL of culture for micro-aerobic conditions to compare the behavior of the different LAB strains. Results are shown in Figure 1. *Lactobacillus delbrueckii* ssp. *bulgaricus* CECT 4005 and CECT 5038 did not produce a significant amount of lactic acid while *L. delbrueckii* ssp. *bulgaricus* CECT 5036 produced up to 14 g L^{-1} of lactic acid racemic mixture in aerobic and micro-aerobic conditions. Furthermore, three strains, *L. delbrueckii* ssp. *bulgaricus* CECT 4006, CECT 5035 and CECT 5037 transformed sugars into lactic acid in micro-aerobic condition with D-LA enantiomeric excess in the same way as *L. delbrueckii* ssp. *delbrueckii* CECT 286. Those strains produced around 15 g L^{-1} of lactic acid with around 75% (e.e.) of D-LA while *L. delbrueckii* ssp. *delbrueckii* CECT 286 reached 92% (e.e.) of D-LA. Therefore, those three strains were selected to study D-LA production from OPW hydrolysate in micro-aerobic conditions.

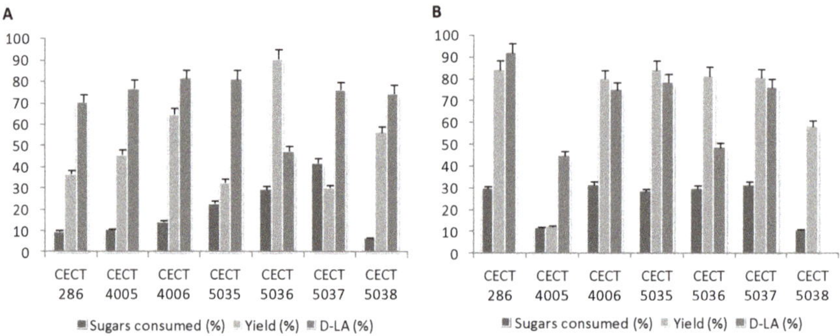

Figure 1. D-LA production in 15 mL tubes with MRS medium containing sugars resembling OPW hydrolysate using LAB strains selected for screening. A. Aerobic conditions. B. Micro-aerobic conditions.

Previous reports showed that lactose rather than glucose markedly increases the growth rate of *L. delbrueckii* ssp. *bulgaricus* strains [33,34]. Therefore, transport systems of sugars other than lactose are likely to vary among these strains and hence, some strains, such as *L. delbrueckii* ssp. *bulgaricus* CECT 4005 and CECT 5038, appear to have difficulties to assimilate the sugars tested in this work. Moreover, strains such as *L. delbrueckii* ssp. *bulgaricus* CECT 5035 and CECT 5037 show low yield in assays at aerobic conditions in the same way as *L. delbrueckii* ssp. *delbrueckii* CECT 286. It is known that during growth, toxic oxygen derivatives are produced for LAB strains in aerobic conditions, but the enzymes required to eliminate them seem not to be expressed in some *L. delbrueckii* ssp. *bulgaricus* strains [42]. Reducing agents may provide protection against toxic products, particularly if growth conditions are not strictly anaerobic. However, with exception of *L. delbrueckii* ssp. *bulgaricus* CECT 5036, the other strains showed higher selectivity to D-LA than *L. delbrueckii* ssp. *delbrueckii* CECT 286 in aerobic conditions and as mentioned above, *L. delbrueckii* ssp. *bulgaricus* CECT 5036 have similar results at aerobic and micro-anaerobic conditions but produced racemic mixture in both cases. *L. delbrueckii* ssp. *bulgaricus* CECT 4005 appears to prefer aerobic conditions but yields are still low.

3.2. Use of OPW Hydrolysate for D-LA Production by Selected Strains

The OPW hydrolysates were prepared following the methodology described in Section 2.4. and developed by de la Torre and colleagues [22] obtaining a glucose yield around 60% *w/w* which corresponds to around 30 g L^{-1}, and obtaining a total sugar concentration above 50 g L^{-1}. Therefore, OPW is a good source of several monosaccharides but also have essential oils rich in limonene and containing terpenes and phenolics with some antimicrobial activity [21]. The tolerance of the strains to the substrate was tested with different concentrations of OPW hydrolysate ranging from 50% to 100% *v/v* diluted with MRS broth. Growth monitoring was performed in a micro-plate incubator for 48 h (Figure 2). Microorganisms grew up well at 50% *v/v* hydrolysate content, but the strain *L. delbrueckii* ssp. *delbrueckii* CECT 286 tolerated the hydrolysate and was able to grow even when hydrolysate content was 100% *v/v*. *Lactobacillus delbrueckii* ssp. *bulgaricus* CECT 4006 appears to be more sensitive to OPW hydrolysate while *L. delbrueckii* ssp. *bulgaricus* CECT 5037 was able to grow up at any OPW concentration; however, the higher the hydrolysate concentration, the higher the lag phase and the lower the growth. Differences lied on the performance of the strains, which is slightly lower when using OPW hydrolysates, probably due to the presence of essential oil components, either terpenes or phenolics. However, *Lactobacilli* are able to withstand relatively high concentrations of citrus extracts [43].

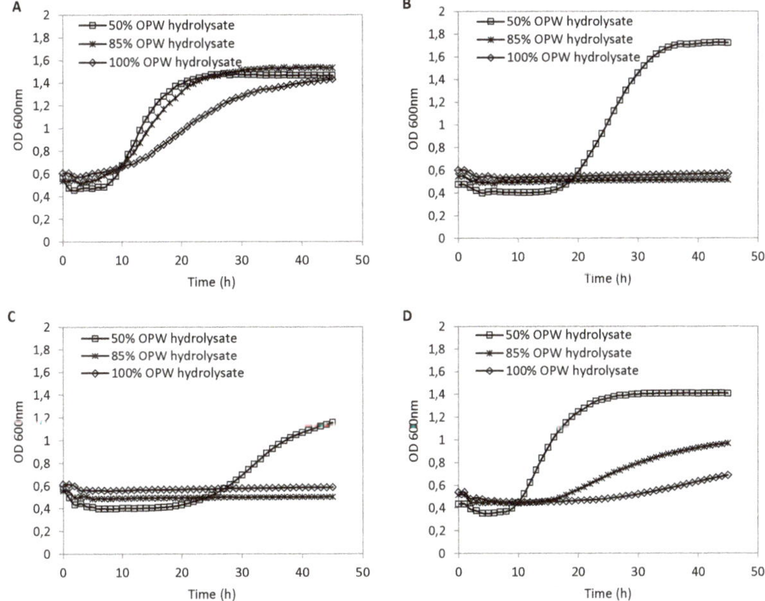

Figure 2. Growth curves for tolerance assays to OPW hydrolysate in microplates and microarebic conditions. (**A**) *Lactobacillus delbrueckii* ssp. *delbreckii* CECT 286. (**B**) *Lactobacillus delbrueckii* ssp. *bulgaricus* CECT 4006. (**C**) *Lactobacillus delbrueckii* ssp. *bulgaricus* CECT 5035. (**D**) *Lactobacillus delbrueckii* ssp. *bulgaricus* CECT 5037. The results were obtained as the average of three replicates and standard deviation was lower than 0.5%.

Concerning the nutritional requirements, previous studies showed that niacin, calcium pantothenate, riboflavin, and vitamin B12 were essential for the growth of *L. delbrueckii* ssp. *bulgaricus*, and that folic acid, pyridoxal, and CaCl$_2$ were important for efficient growth [44,45]. There could be discrepancies due to differences in medium composition or to strain-specific requirements as in the case of *L. delbrueckii* ssp. *bulgaricus* CECT 5037, which not only seems to tolerate hydrolysate, but also seems to grow with less strict nutritional requirements. Although *L. delbrueckii* ssp. *delbueckii* CECT 286 and *L. delbrueckii* ssp. *bulgaricus* CECT 5037 have shown highest robustness cultured in OPW hydrolysate, the next assays were performed using the four selected strains and inoculating the cells recovered from 15% *v/v* of preculture with respect to the volume of culture at 85% *v/v* OPW hydrolysate diluted with MRS medium and micro-aerobic conditions. The inoculum amount was increased to compare the performance of the selected strains with the maximum concentration of OPW hydrolysate during the preliminary fermentation trials.

The optimal growth temperature for *Lactobacilli* ranges from 30 to 40 °C, although some thermophilic strains grow well and have highly activated metabolism at temperatures around 45 °C [35]. The four *Lactobacillus* strains selected were cultured at 37 °C and 45 °C during 120 h to test their activity at conditions as close as possible to those of hydrolysis stage and therefore, to evaluate if the hydrolysis and fermentation stages could be done simultaneously (SSF) as a preliminary result for the future optimization and scale-up of the process. In general, the SSF process offers better yields because it avoids product inhibition and results in higher productivity [10]. Aghababaie and colleagues [36] reported that optimum temperature and pH for growth and lactate production from whey for *L. delbrueckii* ssp. *bulgaricus* were 44 °C and 5.7, respectively. However, the results in Figure 3 show that the strains selected in this study produced D-LA up to 90% (e.e.) in all cases, but the performance of the strains was still better at 37 °C using OPW hydrolysates.

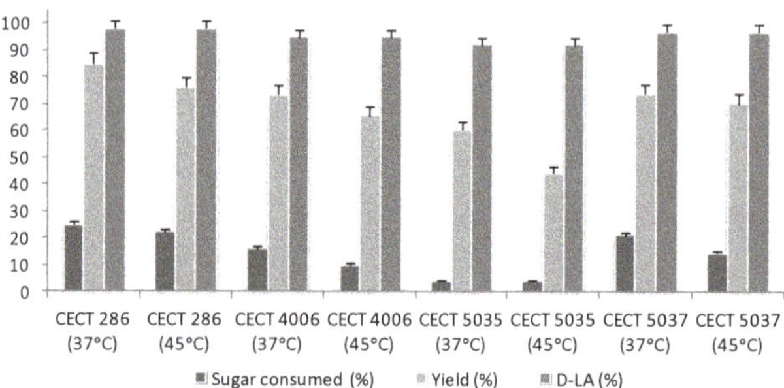

Figure 3. D-LA production results of three *L. delbrueckii* ssp. *bulgaricus* selected in front of *L. delbrueckii* ssp. *delbrueckii* CECT 286 using OPW hydrolysate at 85% *v/v* and incubated at 37 °C and 45 °C to compare strains performance at different temperatures.

Similarly to temperature, the effect of pH change on growth characteristics varied between different species of LAB and in most cases, a decrease of lactate production with a decrease of pH were observed [35]. Therefore, the strains were cultured in 85% *v/v* OPW hydrolysate and pH was adjusted to 5.8 each 24 h during fermentation to test their capacity of production with pH regulation. Cultures were incubated at 37 °C and micro-aerobiosis for 120 h. The results show that sugar consumption and yields were higher when pH was adjusted, and D-LA up to 95% (e.e.) was produced (Figure 4). *L. delbrueckii* ssp. *bulgaricus* CECT 5037 showed the best results in comparison to the other *L. delbrueckii* ssp. *bulgaricus* strains and its performance was comparable to *L. delbrueckii* ssp. *delbrueckii* CECT 286 strain using OPW hydrolysate, whose productivities were between 0.23 and 0.29 g L^{-1} h^{-1}, respectively. Due to the homofermentation of *L. delbrueckii* ssp. *delbrueckii* and *L. delbrueckii* ssp. *bulgaricus* [9,11], only lactic acid could be produced. Nevertheless, a small increase in ethanol concentration onwards of 48 h of fermentation was observed during pH regulation trials. The explanation for this fact, according to the literature [38,46], is that some homofermenters, when grown in limited sugar environment or in the presence of different sugars, can lead to other end products. The main difference is in pyruvate metabolism, but the homofermentation pathway is still used. Additionally, the accumulation of ethanol in the medium (2–3 g L^{-1}) was by far very low to change significantly the generation of the target product. Thus, D-LA continues to be the major fermentation product, and the metabolism of the strains can be considered homofermentative.

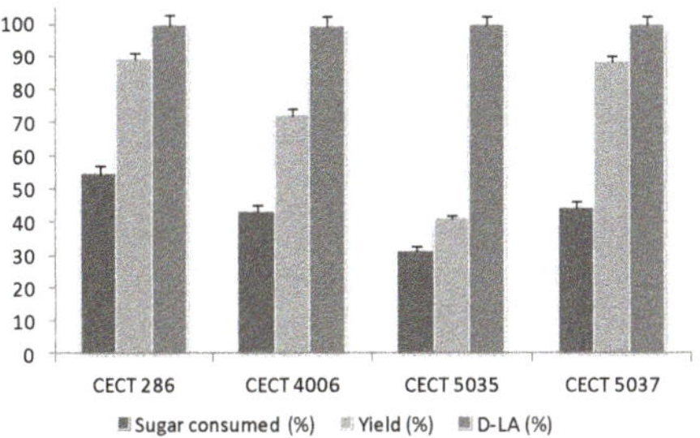

Figure 4. D-LA production results of three *L. delbrueckii* ssp. *bulgaricus* selected in front of *L. delbrueckii* ssp. *delbrueckii* CECT 286 using OPW hydrolysate at 85% *v/v* and adjusting pH at 5.8 each 24 h to evaluate strains performance with pH regulation. The results of standard deviation for the strains with respect to CECT 286 strain are: SD$_{CECT4006}$ = 12.02; SD$_{CECT5035}$ = 34.22; SD$_{CECT5037}$ = 0.69.

3.3. D-LA Production by L. delbrueckii ssp. delbrueckii CECT 286 vs. L. delbrueckii ssp. bulgaricus CECT 5037

Preliminary scale-up assays were performed in 1.5 liter bioreactor by controlling pH at 5.8 in batch mode. Previous results showed that the performance of the strains was better under micro-aerobic conditions, so the bioreactor tests were performed under anaerobic conditions using a nitrogen stream. Cells from 15% *v/v* MRS preculture were inoculated in 85% *v/v* OPW hydrolysate with MRS and supplemented with 5 g L^{-1} of meat extract. According to literature, the more supplemented the medium, the higher the value of final biomass and the higher the productivity of the lactic acid attainable [45,47]. Previous work showed the importance of meat extract and yeast extract in the production of D-LA, probably not due to the total amount of nitrogen but to the growth factors and vitamins contained in these extracts [32]. Fermentation was finished at 72 h (Figure 5), *L. delbrueckii* ssp. *delbrueckii* CECT 286 produced 45 g L^{-1} of lactic acid (99.5% D-LA (e.e.)) with a yield of 86% *w/w* while *L. delbrueckii* ssp. *bulgaricus* CECT 5037 produced 39 g L^{-1} of lactic acid (99.3% D-LA (e.e.)) with a yield of 84% *w/w*.

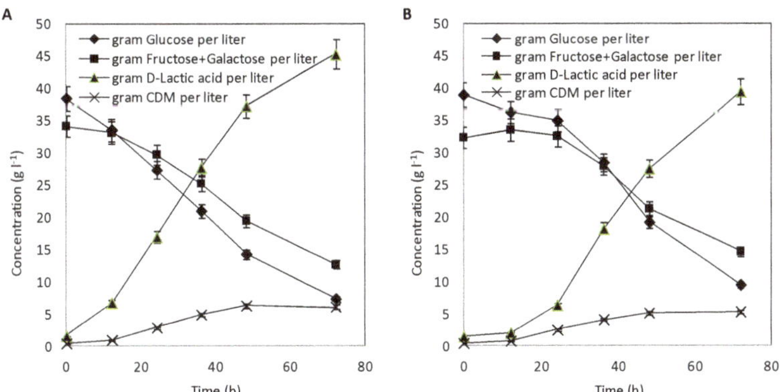

Figure 5. Growth, sugar consumption and D-LA production from OPW hydrolysate in bioreactor and batch mode. A. *Lactobacillus delbrueckii* ssp. *delbreckii* CECT 286. B. *Lactobacillus delbrueckii* ssp. *bulgaricus* CECT 5037. CDM = Cell dry weight.

The yields obtained were similar to those obtained in previous assays by adjusting pH, but the productivities were higher in this case, with values of 0.63 and 0.55 g L^{-1} h^{-1}, respectively. The experiments show that sugars are not completely consumed during fermentation, probably due to deficiencies in the nutritional requirements of the strains. Therefore, the D-LA production process is further optimizable using *L. delbrueckii* ssp. *bulgaricus* CECT 5037 as a promising D-LA producer from OPW hyrolysate and other sustainable feedstocks to contribute in the development of bio-waste refineries. In this regard, commercially important LA-producing LAB strains, such as *Lactobacillus* and *Sporolactobacillus* strains, are particularly useful because of their high lactic acid yield, high acid tolerance, and their ability to be metabolically engineered [9,12]. Efficient conversion of biomass to D-LA still faces considerable challenges, such as high energy demand and high enzyme cost for pretreatment of lignocellulosic biomass, inefficiency of sugar utilization by microorganisms, and undesired byproducts generated during the fermentation process [46]. Table 3 summarizes studies of D-LA production from sustainlable feedstocks such as agro-industrial residues by wild-type stains. These results indicate that OPW hydrolysate is an interesting feedstock for the production of D-LA, since the product yield is close to its theoretical value (1 g g^{-1}) in most cases. Apart from that, the productivity value is quite high and very attractive when industrial developments are envisaged [32,48]. It is common that bioprocesses based on biomass waste give poorer results than their control experiments based on sugar mixtures resembling the hydrolysates composition. In this case, yields achieved have close values in both cases. According to the achieved purity of lactic acid (> 95%), differences were not observed when OPW hydrolysate is used, suggesting that the waste compounds do not influence D-LA purity.

Table 3. D-LA production from sustainable feedstocks in batch cultures by wild-type LAB strains.

Feedstock	Microorganism	Process	Yield (g g^{-1})	Productivity (g L^{-1} h^{-1})	D-LA (%)	Reference
Rice starch	*L. delbrueckii* LD 0028	SHF	0.70	1.55	97.5	[49]
Defatted rice bran	*L. delbrueckii* IFO 3202	SSF	0.78	1.25	> 95	[10]
Sugarcane molasses	*L. delbrueckii* JCM 1148	–	0.90	1.48	97.2	[31]
Sugarcane juice	*L. delbrueckii* JCM 1148	–	0.95	1.66	98.3	[31]
Sugar beet juice	*L. delbrueckii* JCM 1148	–	0.88	1.16	97.6	[31]
Microalga	*L. coryniformis* ssp. *torquens* ATCC 25600	SSF	0.46	1.02	95.8	[50]
Curcuma longa waste	*L. coryniformis* ssp. *torquens* ATCC 25600	SSF	0.65	2.08	> 95	[51]
Pulp	*L. delbrueckii* ATCC 9649	SHF	0.83	1.01	99	[9]
Casein whey permeate	*L. delbrueckii* ssp. *lactis* ATCC 4797	–	0.49	0.61	> 98	[52]
Pulp mill residues	*L. coryniformes* ssp. *torquens* ATCC 25600	SHF	0.97	2.80	99	[53]
Orange peel waste	*L. delbrueckii* ssp. *delbrueckii* CECT 286	SHF	0.88	2.35	> 95	[32]
Orange peel waste	*L. delbrueckii* ssp. *delbrueckii* CECT 286	SHF	0.86	0.63	99.5	This study
Orange peel waste	*L. delbrueckii* ssp. *bulgaricus* CECT 5037	SHF	0.84	0.55	99.3	This study *

* Preliminary results of LAB screening for further optimization. SSF: Simultaneous saccharification and fermentation; SHF: Separate hydrolysis and fermentation.

The results obtained with the *L. delbrueckii* ssp. *bulgaricus* CECT 5037 strain are promising since performance of the strain was comparable to *L. delbrueckki* ssp. *delbrueckii* CECT 286 strain performance using OPW hydrolysate at the conditions tested in this work. Previous studies show that the reference strain can reach a productivity of 2.35 g L^{-1} h^{-1} when fermentation conditions are optimized [32]. Therefore, future work with *L. delbrueckii* ssp. *bulgaricus* CECT 5037 will be focused on optimization of fermentation methodology, including the method of inoculation of the cultures, improvement of culture media by testing low cost nutrient sources, as well as the evaluation of operational costs in developing a sustainable lactic acid production process.

4. Conclusions

Six strains of the species *Lactobacillus delbrueckii* ssp. *bulgaricus* were evaluated for the production of D-LA from OPW hydrolysate in comparison to the reference *Lactobacillus delbrueckii* ssp. *delbrueckii* CECT 286 strain. Remarkably, *Lactobacillus delbrueckii* ssp. *bulgaricus* CECT 5037 is able to tolerate the OPW hydrolysate and produce D-LA up to 95% (e.e.). The results of strain performance show a yield of 84% *w/w* for lactic acid production that is close to the yield of 86% *w/w* obtained with the reference *Lactobacillus delbrueckii* ssp. *delbrueckii* CECT 286 strain in this work and 88% *w/w* reported from previous works when process improvement was foreseen. Experiments will be underway to develop the process and further optimization will contribute to providing a suitable alternative to biowaste-refinery processes using OPW and other residual feedstocks as a potential substrate for valorisation.

Author Contributions: D.R. and M.T. conceived the research; A.R. and D.B. designed the experiments; D.B. performed the experiments; A.R. evaluated the results; D.B. wrote the paper. All authors have read and agreed to the published version of the manuscript.

Funding: This research was funded by European Commission, grant number EIB.12.007

Conflicts of Interest: The authors declare no conflict of interest. The funders had no role in the design of the study; in the collection, analyses, or interpretation of data; in the writing of the manuscript, or in the decision to publish the results.

References

1. Esteban, J.; Ladero, M. Food waste as a source of value-added chemicals and materials: A biorefinery perspective. *Int. J. Food. Sci. Technol.* **2018**, *53*, 1095–1108. [CrossRef]
2. John, R.P.; Anisha, G.; Nampoothiri, K.M.; Pandey, A. Direct lactic acid fermentation: Focus on simultaneous saccharification and lactic acid production. *Biotechnol. Adv.* **2009**, *27*, 145–152. [CrossRef]
3. Abdel-Rahman, M.A.; Tashiro, Y.; Sonomoto, K. Lactic acid production from lignocellulose-derived sugars using lactic acid bacteria: Overview and limits. *J. Biotechnol.* **2011**, *156*, 286–301. [CrossRef]
4. Klotz, S.; Kaufmann, N.; Kuenz, A.; Prüße, U. Biotechnological production of enantiomerically pure D-lactic acid. *Appl. Microbiol. Biotechnol.* **2016**, *100*, 9423–9437. [CrossRef]
5. Wang, Y.; Tashiro, Y.; Sonomoto, K. Fermentative production of lactic acid from renewable materials: Recent achievements, prospects, and limits. *J. Biosci. Bioeng.* **2015**, *119*, 10–18. [CrossRef]
6. Singhvi, M.; Zendo, T.; Sonomoto, K. Free lactic acid production under acidic conditions by lactic acid bacteria strains: Challenges and future prospects. *Appl. Microbiol. Biotechnol.* **2018**, *102*, 1–14. [CrossRef]
7. John, R.P.; Nampoothiri, K.M.; Pandey, A. Fermentative production of lactic acid from biomass: An overview on process developments and future perspectives. *Appl. Microbiol. Biotechnol.* **2007**, *74*, 524–534. [CrossRef]
8. Tashiro, Y.; Kaneko, W.; Sun, Y.; Shibata, K.; Inokuma, K.; Zendo, T.; Sonomoto, K. Continuous D-lactic acid production by a novel thermo tolerant *Lactobacillus delbrueckii* subsp. lactis QU 41. *Appl. Microbiol. Biotechnol.* **2011**, *89*, 1741–1750. [CrossRef]
9. Zhang, Y.; Vadlani, P.V. D-Lactic acid biosynthesis from biomass derived sugars via *Lactobacillus delbrueckii* fermentation. *Bioprocess Biosyst. Eng.* **2013**, *36*, 1897–1904. [CrossRef]
10. Tanaka, T.; Hoshina, M.; Tanabe, S.; Sakai, K.; Ohtsubo, S.; Taniguchi, M. Production of D-lactid acid from defatted rice bran by simultaneous saccarification and fermentation. *Bioresour. Technol.* **2006**, *97*, 211–217. [CrossRef]
11. Abdel-Rahman, M.A.; Tashiro, Y.; Sonomoto, K. Recent advances in lactic acid production by microbial fermentation processes. *Biotechnol. Adv.* **2013**, *31*, 877–902. [CrossRef]
12. Zhang, Y.; Vadlani, P.V.; Kumar, A.; Hardwidge, P.R.; Govind, R.; Tanaka, T.; Kondo, A. Enhanced D-lactic acid production from renewable resources using engineered *Lactobacillus plantarum*. *Appl. Microbiol. Biotechnol.* **2016**, *100*, 279–288. [CrossRef]
13. Idris, A.; Suzana, W. Effect of sodium alginate concentration, bead diameter, initial pH and temperature on lactic acid production from pineapple waste using immobilized *Lactobacillus delbrueckii*. *Process. Biochem.* **2006**, *41*, 1117–1123. [CrossRef]
14. Choi, M.; Al-Zahrani, S.M.; Lee, S.Y. Kinetic model-based feed-forward controlled fed-batch fermentation of *Lactobacillus rhamnosus* for the production of lactic acid from Arabic date juice. *Bioprocess Biosyst. Eng.* **2014**, *37*, 1007–1015. [CrossRef]
15. Hu, J.; Zhang, Z.; Lin, Y.; Zhao, S.; Mei, Y.; Liang, Y.; Peng, N. High-titer lactic acid production from NaOH-pretreated corn stover by *Bacillus coagulans* LA204 using fed-batch simultaneous saccharification and fermentation under non-sterile condition. *Bioresour. Technol.* **2015**, *182*, 251–257. [CrossRef]

16. Hama, S.; Mizuno, S.; Kihara, M.; Tanaka, T.; Ogino, C.; Noda, H.; Kondo, A. Production of D-lactic acid from hard wood pulp by mechanical milling followed by simultaneous saccharification and fermentation using metabolically engineered *Lactobacillus plantarum*. *Bioresour. Technol.* **2015**, *187*, 167–172. [CrossRef]

17. Okano, K.; Hama, S.; Kihara, M.; Noda, H.; Tanaka, T.; Kondo, A. Production of optically pure D-lactic acid from brown rice using metabolically engineered Lactobacillus plantarum. *Appl Microbiol. Biotechnol.* **2017**, *101*, 1869–1875.

18. Van Dyk, J.S.; Gama, R.; Morrison, D.; Swart, S.; Pletschke, B.I. Food processing waste: Problems, current management and prospects for utilisation of the lignocellulose component through enzyme synergistic degradation. *Renew. Sust. Energ. Rev.* **2013**, *26*, 521–531. [CrossRef]

19. Girotto, F.; Alibardi, L.; Cossu, R. Food waste generation and industrial uses: A review. *Waste. Manag.* **2015**, *45*, 32–41. [CrossRef]

20. Rezzadori, K.; Benedetti, S.; Amante, E.R. Proposals for the residues recovery: Orange waste as raw material for new products. *Food Bioprod. Process.* **2012**, *90*, 606–614. [CrossRef]

21. Negro, V.; Mancini, G.; Ruggeri, B.; Fino, D. Citrus waste as feedstock for bio-based products recovery: Review on limonene case study and energy valorization. *Bioresour. Technol.* **2016**, *214*, 806–815. [CrossRef]

22. De la Torre, I.; Ravelo, M.; Segarra, S.; Tortajada, M.; Santos, V.E.; Ladero, M. Study on the effects of several operational variables on the enzymatic batch saccharification of orange solid waste. *Bioresour. Technol.* **2017**, *245*, 906–915. [CrossRef]

23. Choi, I.S.; Lee, Y.G.; Khanal, S.K.; Park, B.J.; Bae, H.-J. A low-energy, cost-effective approach to fruit and citrus peel waste processing for bioethanol production. *Appl. Energy* **2015**, *140*, 65–74. [CrossRef]

24. Rafiq, S.; Kaul, R.; Sofi, S.A.; Bashir, N.; Nazir, F.; Ahmad Nayik, G. Citrus peel as a source of functional ingredient: A review. *J. Saudi Soc. Agric. Sci.* **2016**, *17*, 351–358. [CrossRef]

25. Martín, M.A.; Siles, J.A.; Chica, A.F.; Martín, A. Biomethanization of orange peel waste. *Bioresour. Technol.* **2010**, *101*, 8993–8999. [CrossRef]

26. Pourbafrani, M.; Forgács, G.; Horváth, I.S.; Niklasson, C.; Taherzadeh, M.J. Production of biofuels, limonene and pectin from citrus wastes. *Bioresour. Technol.* **2010**, *101*, 4246–4250. [CrossRef]

27. Oberoi, H.S.; Vadlani, P.V.; Madl, R.L.; Saida, L.; Abeykoon, J.P. Ethanol production from orange peels: Two-stage hydrolysis and fermentation studies using optimized parameters through experimental design. *J. Agric. Food Chem.* **2010**, *58*, 3422–3429. [CrossRef]

28. Lugo-Lugo, V.; Barrera-Díaz, C.; Ureña-Núñez, F.; Bilyeu, B.; Linares-Hernández, I. Biosorption of Cr(III) and Fe(III) in single and binary systems onto pretreated orange peel. *J. Environ. Manag.* **2012**, *112*, 120–127. [CrossRef]

29. Mazzoli, R.; Bosco, F.; Mizrahi, I.; Bayer, E.A.; Pessione, E. Towards lactic acid bacteria-based biorefineries. *Biotechnol. Adv.* **2014**, *32*, 1216–1236. [CrossRef]

30. Eiteman, M.A.; Ramalingam, S. Microbial production of lactic acid. *Biotechnol. Lett.* **2015**, *37*, 955–972. [CrossRef]

31. Calabia, B.P.; Tokiwa, Y. Production of D-lactic acid from sugar cane molasses, sugarcane juice and sugar beet juice by *Lactobacillus delbrueckii*. *Biotechnol. Lett.* **2007**, *29*, 1329–1332. [CrossRef]

32. De la Torre, I.; Ladero, M.; Santos, V.E. Production of D-lactic acid by *Lactobacillus delbrueckii* ssp. delbrueckii from orange peel waste: Techno-economical assesment of nitrogen sources. *Appl. Microbiol. Biotechnol.* **2018**, *102*, 10511–10521.

33. Benthin, S.; Villadsen, J. Production of optically pure D-lactate by *Lactobacillus bulgaricus* and purification by crystallisation and liquid/liquid extraction. *Appl. Microbiol. Biotechnol.* **1995**, *42*, 826–829. [CrossRef]

34. Benthin, S.; Villadsen, J. Different inhibition of *Lactobacillus delbrueckii* subsp. bulgaricus by D- and L-lactic acid: Effects on lag phase, growth rate and cell yield. *J. Appl. Biotecnol.* **1995**, *78*, 647–654.

35. Adamberg, K.; Kaska, S.; Lahta, T.M.; Paalmea, T. The effect of temperature and pH on the growth of lactic acid bacteria: A pH-auxostat study. *Int. J. Food. Microbiol.* **2003**, *85*, 171–183. [CrossRef]

36. Aghababaie, M.; Beheshti, M.; Khanahmadi, M. Effect of temperature and pH on formulating the kinetic growth parameters and lactic acid production of *Lactobacillus bulgaricus*. *Nutr. Food. Sci. Res.* **2014**, *1*, 49–56.

37. Bernard, N.; Ferain, T.; Garmyn, D.; Hols, P.; Delcour, J. Cloning of the D-lactate dehydrogenase gene from *Lactobacillus delbrueckii* subsp. bulgaricus by complementation in Escherichia coli. *FEBS Lett.* **1991**, *290*, 61–64. [CrossRef]

38. Hofvendahl, K.; Hahn–Hägerdal, B. Factors affecting the fermentative lactic acid production from renewable resources. *Enzyme Microb. Technol.* **2000**, *26*, 87–107. [CrossRef]

39. Sluiter, A.; Ruiz, R.; Scarlata, C.; Sluiter, J.; Templeton, D. *Determination of Extractives in Biomass*; Laboratory Analytical Procedure (LAP); Sluiter, A., Ed.; National Renewable Energy Laboratory: Golden, CO, USA, 2008.

40. Sluiter, A.; Hames, B.; Ruiz, R.; Scarlata, C.; Sluiter, J.; Templeton, D. *Determination of Sugars, Byproducts, and Degradation Products in Liquid Fraction Process Samples*; Laboratory Analytical Procedure (LAP); Sluiter, A., Ed.; National Renewable Energy Laboratory: Golden, CO, USA, 2008.

41. Sluiter, A.; Hames, B.; Ruiz, R.; Scarlata, C.; Sluiter, J.; Templeton, D.; Crocker, D. *Determination of Structural Carbohydrates and Lignin in Biomass*; Laboratory Analytical Procedure (LAP); Sluiter, A., Ed.; National Renewable Energy Laboratory: Golden, CO, USA, 2012.

42. Archibald, F.S.; Fridovich, I. Manganese and defenses against oxygen toxicity in *Lactobacillus plantarum*. *J. Bacteriol.* **1981**, *145*, 442–451.

43. Bevilacqua, A.; Corbo, M.R.; Sinigaglia, M. In vitro evaluationof the antimicrobial activity of eugenol, limonene, and citrus extract against bacteriaand yeasts, representativeofthe spoilingmicroflora of fruit juices. *J. Food Prot.* **2010**, *73*, 888–894. [CrossRef]

44. Grobben, G.J.; Chin-Joe, I.; Kitzen, V.A.; Boels, I.C.; Boer, F.; Sikkema, J.; Smith, M.R.; De Bont, J.A.M. Enhancement of exopolysaccharide production by *Lactobacillus delbrueckii* subsp. bulgaricus NCFB 2772 with a simplified defined medium. *Appl. Environ. Microbiol.* **1998**, *64*, 1333–1337.

45. Chervaux, C.; Ehrlich, S.D.; Maguin, E. Physiological Study of *Lactobacillus delbrueckii* subsp. bulgaricus Strains in a Novel Chemically Defined Medium. *Appl. Environ. Microbiol.* **2000**, *66*, 5306–5311.

46. Zhang, Y.; Yoshida, M.; Vadlani, P.V. Biosynthesis of D-lactic acid from lignocellulosic biomass. *Biotechnol. Lett.* **2018**, *40*, 1167–1179. [CrossRef]

47. Kwon, S.; Lee, P.C.; Lee, E.G.; Chang, Y.K.; Chang, N. Production of lactic acid by *Lactobacillus rhamnosus* with vitamin-supplemented soybean hydrolysate. *Enzym. Microb. Technol.* **2000**, *26*, 209–215. [CrossRef]

48. De la Torre, I.; Acedos, M.G.; Ladero, M.; Santos, V.E. On the use of resting *L. delbrueckii* spp. delbrueckii cells for D-lactic acid production from orange peel wastes hydrolysates. *Biochem. Eng. J.* **2019**, *145*, 162–169.

49. Fukushima, K.; Sogo, K.; Miura, S.; Kimura, Y. Production of D-lactic acid by bacterial fermentation of rice starch. *Macromol. Biosci.* **2004**, *4*, 1021–1027. [CrossRef]

50. Nguyen, C.M.; Kim, J.S.; Song, J.K.; Choi, G.J.; Choi, Y.H.; Jang, K.S.; Kim, J.C. D-lactic acid production from dry biomass of *Hydrodictyon reticulatum* by simultaneous saccharification and co-fermentation using *Lactobacillus coryniformis* subsp torquens. *Biotechnol. Lett.* **2012**, *34*, 2235–2240. [CrossRef]

51. Nguyen, C.M.; Kim, J.S.; Nguyen, T.N.; Kim, S.K.; Choi, G.J.; Choi, Y.H.; Jang, K.S.; Kim, J.C. Production of L- and D-lactic acid from waste *Curcuma longa* biomass through simultaneous saccharification and cofermentation. *Bioresour. Technol.* **2013**, *146*, 35–43. [CrossRef]

52. Prasad, S.; Srikanth, K.; Limaye, A.M.; Sivaprakasam, S. Homo-fermentative production of d-lactic acid by *Lactobacillus* sp. employing casein whey permeate as a raw feed-stock. *Biotechnol. Lett.* **2014**, *36*, 1303–1307. [CrossRef]

53. De Oliveira Moraes, A.; Ramirez, N.I.B.; Pereira, N. Evaluation of the fermentation potential of pulp mill residue to produce D (−)-lactic acid by separate hydrolysis and fermentation using *Lactobacillus coryniformis* subsp. torquens. *Appl. Biochem. Biotechnol.* **2016**, *180*, 1574–1585. [CrossRef]

 *fermentation*

Article

The Second-Generation Biomethane from Mandarin Orange Peel under Cocultivation with Methanogens and the Armed *Clostridium cellulovorans*

Hisao Tomita [1,†] and Yutaka Tamaru [1,2,3,*]

1   Department of Life Sciences, Graduate School of Bioresources, Mie University, 1577 Kurimamachiya, Tsu, Mie 514-8507, Japan; 516d301@m.mie-u.ac.jp or drhtomitaent@gmail.com
2   Department of Bioinformatics, Institute of Advanced Research Center, Mie University, 1577 Kurimamachiya, Tsu, Mie 514-8507, Japan
3   Research Center of Smart Cell Innovation, Mie University, 1577 Kurimamachiya, Tsu, Mie 514-8507, Japan
*   Correspondence: ytamaru@bio.mie-u.ac.jp; Tel.: +81-59-231-9560
†   Present address: 15 Maeda, Fujie, Higashiura, Chita, Aichi 470-2105, Japan.

Received: 10 September 2019; Accepted: 31 October 2019; Published: 4 November 2019

Abstract: This study demonstrates that the consortium, which consists of the microbial flora of methane production (MFMP) and *Clostridium cellulovorans* grown with cellulose, can perform the direct conversion of cellulosic biomass to methane. The MFMP was taken from a commercial methane fermentation tank and was extremely complicated. Therefore, *C. cellulovorans* grown with cellobiose could not perform high degradation ability on cellulosic biomass due to competition by various microorganisms in MFMP. Focusing on the fact that *C. cellulovorans* was cultivated with cellulose, which is armed with cellulosome, so that it is now armed *C. cellulovorans*; the direct conversion was carried out by the consortium which consisted of MFMP and the armed *C. cellulovorans*. As a result, the consortium of *C. cellulovorans* grown with cellobiose and MFMP (CCeM) could not degrade the purified cellulose and mandarin orange peel. However, MFMP and the armed *C. cellulovorans* reduced 78.4% of the total sugar of the purified cellulose such as MN301, and produced 6.89 mL of methane simultaneously. Furthermore, the consortium consisted of MFMP and the armed *C. cellulovorans* degraded mandarin orange peel without any pretreatments and produced methane that was accounting for 66.2% of the total produced gas.

Keywords: methanogenesis; cellulosic biomass degradation; consortia; armed *C. cellulovorans*

1. Introduction

Although the first-generation biofuels, which are made from corn and sugarcane, have become widespread, there is concern about competition with food supply. Without competing for food, second-generation biofuels are produced from non-edible biomass such as agricultural wastes and cellulosic substrates [1,2].

A plant cell wall is composed of cellulose, hemicellulose, lignin, pectin, etc. Cellulose is a fiber of D-glucose monomers and has strong crystalline [3]. Moreover, cellulose, hemicellulose, and lignin compose the rigid and complex structures [4]. Hemicellulose is a heteropolymer such as xylan, glucuronoxylan, arabinoxylan, glucomannan, and xyloglucan. In addition, lignin, phenol compounds, reinforces the structure of cellulose and hemicellulose, making it more difficult to degrade. Thus, since rigid and complex structures are constructed in cellulosic biomass, it is very difficult to degrade them enzymatically.

Orange juice is one of the major fruit juices and almost the same amount of orange waste as orange juice comes out as a byproduct in orange juice factories. Therefore, it has been considered that such orange wastes are available for non-edible biomass all over the world. However, d-limonene, which is included in citrus, has an extremely toxic effect on fermenting microorganisms [5,6]. Depending on the type of citrus, the orange peel contains approximately 3% limonene. Therefore, it was necessary to separate d-limonene before the cultivation or to protect microorganisms from d-limonene by encapsulation or immobilization [7,8]. Recently, it was reported that *Clostridium cellulovorans* can degrade orange wastes in the culture including d-limonene and *Clostridium beijerinckii* can carry out isopropanol–butanol–ethanol (IBE) fermentation, which is a bacterial fermentation process producing isopropanol instead of acetone on acetone–ethanol–butanol (ABE) fermentation [9–11]. Thus, the breakthrough was obtained utilizing orange wastes for second-generation biofuels without any pre-treatment.

Some *Clostridia*, such as *C. cellulovorans* and *Clostridium thermocellum*, are known to have the ability to degrade cellulosic biomass efficiently using cellulosomes and secreted non-cellulosomal enzymes [12]. Among those species, we have been studying *C. cellulovorans*, which is a mesophilic and anaerobic cellulolytic bacterium [13]. *C. cellulovorans* degrades not only cellulose but also hemicelluloses consisting of xylose, fructose, galactose, and mannose [14–16]. Whole-genome sequencing of *C. cellulovorans* and the exoproteome profiles revealed 57 cellulosomal protein-encoding genes and 168 secreted-carbohydrase-encoding genes [17,18]. Furthermore, the high degradation ability on plant cell walls has so far been reported [19]. *C. cellulovorans* grown in the culture with cellulose has large protuberances on its surface [20] and the protuberances contain cellulosomes [21]; on the other hand, *C. cellulovorans* grown in the culture with cellobiose does not have the protuberances (Figure 1a,b). *C. cellulovorans* acquires a high ability to degrade the plant cell wall with cellulosome, in other words, *C. cellulovorans* is armed with cellulosome to attack the plant cell wall. Therefore, it can be said that it is the armed *C. cellulovorans* that grow with cellulose. Although *C. cellulovorans* has high degradation ability on plant cell walls in the *C. cellulovorans* monoculture, there are challenges in that *C. cellulovorans* cannot perform high degradation ability in the co-culture or consortia of other microorganisms. Most of the reports used *C. cellulovorans* grown with cellobiose, and these challenges were most likely derived using *C. cellulovorans* grown with cellobiose. Therefore, this study focuses on utilizing the armed *C. cellulovorans* especially for the consortium of other microorganisms.

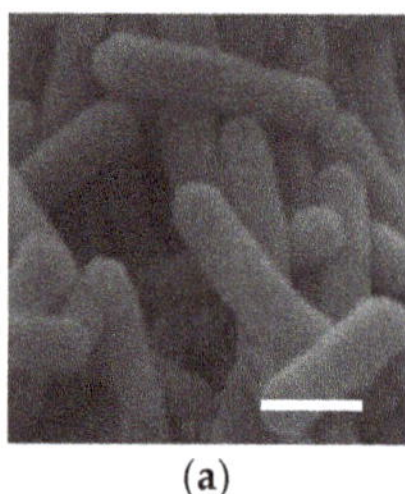

(a)

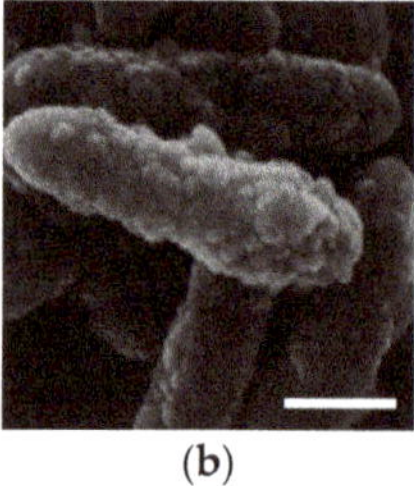

(b)

Figure 1. (**a**) Cell surface of *C. cellulovorans* grown with cellobiose and (**b**) cell surface of *C. cellulovorans* grown with cellulose. White bar indicates 2 μm. Modified from [21].

Many studies on methane (CH_4) fermentation have been reported in a wide range of study fields [22]. Methane fermentation using cellulosic biomass is also second-generation biomethane. Since methane production is carried out by the complex microbial flora including methanogens, the second-generation biomethane process needs the consortium to be constructed with microbial flora of methane production (MFMP) and microorganisms which can degrade cellulosic biomass, such as *C. cellulovorans*. The consortium of *C. cellulovorans* with MFMP (CCeM) can degrade sugar beet pulp, which is the residue in a sugar refinery factory, and ferments biogas included methane simultaneously [23]. However, the relict sugars in sugar beet pulp were possible to help *C. cellulovorans* to survive and coexist with MFMP, and the same CCeM could not carry out degrading the purified cellulose and producing methane.

In the present study, we investigated the degradation ability on cellulosic biomass of CCeM that was consistent with the armed *C. cellulovorans* (ACCeM).

2. Materials and Methods

2.1. Materials

Mandolin oranges were purchased at a grocery store in Japan in 2017. Flavedo and albedo, hereafter called removed peel, were cut into strips with scissors just after removing from a mandolin orange. The strip size was approximately 10 mm in length and 2 mm in width (Figure 2). The peel was used as fresh, not dried, and ground. Furthermore, the peel was not treated by any chemicals. The dried weight of the removed peel was measured, it contained 71.6% water. The substrate concentration of removed peel, the purified celluloses, such as Avicel (Sigma, St. Louis, MO, USA) and MN301 (MACHEREY-NAGEL, Düren, Deutschland), and cellobiose (Sigma, St. Louis, MO, USA) was 0.5% (*w/v*) of the dry weight. Avicel is crystalline cellulose powder that is industrially refined from natural cellulose, and the particle size of Avicel is less than 50 μm. MN301 is also industrially refined cellulose powder, and 80% of the particle size is less than 160 μm.

Figure 2. (**a**) Removed peel before cutting. (**b**) Strips of removed peel in the medium.

2.2. Microorganism and Culture Condition

The medium was partially modified by *Clostridium cellulovorans* medium [11]. One litter medium contained 4 g of yeast extract, 1 mg of Resazurin salt, 1 g of L-cysteine-HCl, 5 g of $NaHCO_3$, 0.45 g of K_2HPO_4, 0.45 g of KH_2PO_4, 0.3675 g of NH_4Cl, 0.9 g of NaCl, 0.1575 g of $MgCl_2.6H_2O$, 0.12 g of $CaCl_2.2H_2O$, 0.85 mg of $MnCl_2.4H_2O$, 0.942 mg of $CoCl_2.6H_2O$, 5.2 mg of Na_2EDTA, 1.5 mg of $FeCl_2.4H_2O$, 0.07 mg of $ZnCl_2$, 0.1 mg H_3BO_3, 0.017 mg of $CuCl_2.2H_2O$, 0.024 mg of $NiCl_2.6H_2O$, 0.036 mg of $Na_2MoO_4.2H_2O$, 6.6 mg of $FeSO_4.7H_2O$, and 0.1 g of p-aminobenzoic acid and was adjusted to pH 7. *C. cellulovorans* 743B (ATCC 35296) was used and anaerobically cultivated in 0.5% (*w/v*) cellobiose, Avicel and MN301 at 37 °C stationary for 19 h, for 4 days and 2 days respectively. The MFMP was obtained from methane fermentation digested liquid in January 2017 at Gifu in Japan. The MFMP was anaerobically cultivated in *Clostridium cellulovorans* medium with 0.5% (*w/v*) glucose (Wako) and 0.25% (*w/v*) cellobiose, or 0.5% (*w/v*) ryegrass leaves at 37 °C for 19 h stationary. The ryegrass leaves were obtained in May 2017 at Aichi in Japan, and were dried and grained to a powder.

2.3. Data Deposition

The sequences of MFMP that are reported in this paper have been deposited in the DNA Data Bank of Japan (DDBJ) (accession 104 no. DRR160954).

2.4. Measurement of Total Sugar Concentration

The total sugar concentration was measured from the precipitation after centrifugation which was 20,000 rpm at 4 °C for 5 min. The precipitation was washed by phosphate buffered salts and 5N NaOH. The total sugar concentration of the washed precipitation was measured by the phenol-sulfuric acid method as D-glucose equivalents.

2.5. Gas Concentration

The produced gas volume of the head space in the culture vial was collected and measured by a syringe (Terumo, Tokyo, Japan). The concentrations of CH_4, H_2, and CO_2 were measured by a gas chromatograph GC-8A (Shimadzu, Kyoto, Japan) with a Thermal Conductivity Detector (TCD) and a column SINCARBON ST (full length 6 m, inner diameter 3 mm; Shinwa, Kyoto, Japan). The column temperature was 200 °C. Argon was a carrier gas and set at a flow rate of 50 mL/min. The injection volume of each sample was 5 mL. The concentration of CH_4 was also measured by a gas chromatograph GC-2010plus (Shimadzu, Kyoto, Japan) with a capillary column Rt-Q-BOND (30 m, inner diameter. 0.32 mm; RESTEK, Centre, PA, USA). The oven temperature was 250 °C and the column temperature was 150 °C. Nitrogen was the carrier gas and set at a flow rate of 1.21 mL/min. The injection volume of each sample was 0.5 mL.

2.6. Organic Acid Concentration

The concentration of organic acids was measured by high-performance liquid chromatography (HPLC), CBM-20A, LC-20AD, CTO-20AC, SPD-20A, and DGU-20A3 (Shimadzu, Kyoto, Japan) with a UV detector and a column KC-811 (300 mm × 2 mm, inner diameter. 8 mm; Showa Denko, Tokyo, Japan). The column temperature was 60 °C. The Bromothymol blue (BTB) post-column method was used. The eluent was 2 mM perchloric acid, and the flow rate was 1.0 mL/min. The reagent was 0.2 mM BTB and 15 mM disodium hydrogen phosphate, and the flow rate was 1.2 mL/min at the wavelength of 445 nm. The injection volume of each sample was 20 µL.

2.7. Cell Growth

Cell growth was measured by Lumitester PD-20, LuciPac Pen and adenosine triphosphate (ATP) eliminating enzyme (Kikkoman Biochemifa, Tokyo, Japan). It is known that integrated intracellular ATP concentration correlates with cell growth [24]. Cell growth was estimated by measuring ATP concentration of 0.1 mL of cell culture according to the manufacturer's instruction and was expressed by the relative light unit (RLU) value.

2.8. Statistics

The data were analyzed for statistical significance using Welch's *t*-test. The difference was assessed with a two-sided test with an α level of 0.05.

3. Results

3.1. Degrading Cellulose and Removed Peel under Cocultivation with Methanogens and Non-Armed C. cellulovorans

C. cellulovorans was pre-cultivated in media containing 0.5% (*w/v*) of cellobiose, which meant *C. cellulovorans* was not armed with cellulosome. Anaerobic batch cultivations of *C. cellulovorans*, CCeM, and MFMP were carried out in a 50-mL medium containing 0.5% (*w/v*) of Avicel and removed peel at 37 °C without shaking.

According to measured cell growth of precultures, inoculation amounts for monocultures of *C. cellulovorans* and MFMP were decided so that initial RLU values of each monoculture came close to 2000. RLU values of the preculture of *C. cellulovorans* and MFMP were 45,310 and 50,494, respectively. The inoculation volume was decided as 2 mL for 50 mL monoculture; it was 26 times the dilution so that the initial RLU value of monoculture of *C. cellulovorans* and MFMP were 1743 and 1942, respectively. CCeM was inoculated 2 mL each from both precultures so that concentrations of cell growth against substrate became the same as the monoculture.

RLU values increased for 1 day of cultivation and then cell growth was observed in all cultures (Figure 3a,b). Interestingly, RLU profiles of CCeM and MFMP were quite similar, even CCeM included *C. cellulovorans* and MFMP both. The total sugar concentrations were significantly reduced in *C. cellulovorans* monocultures with Avicel and removed peel after 11 days of cultivation, it was demonstrated that *C. cellulovorans* could degrade Avicel and removed peel. However, the total sugar concentrations in CCeM with Avicel and removed peel were almost the same as that of MFMP; *C. cellulovorans* could not degrade Avicel and removed peel in cultures included MFMP (Figure 3c,d). It was thought that *C. cellulovorans* did not grow in CCeM because cell growths of CCeM and MFMP were almost the same. Methane in the head space was not detected in CCeM with removed peel and that was slightly detected in CCeM and MFMP with Avicel (Figure 3e,f).

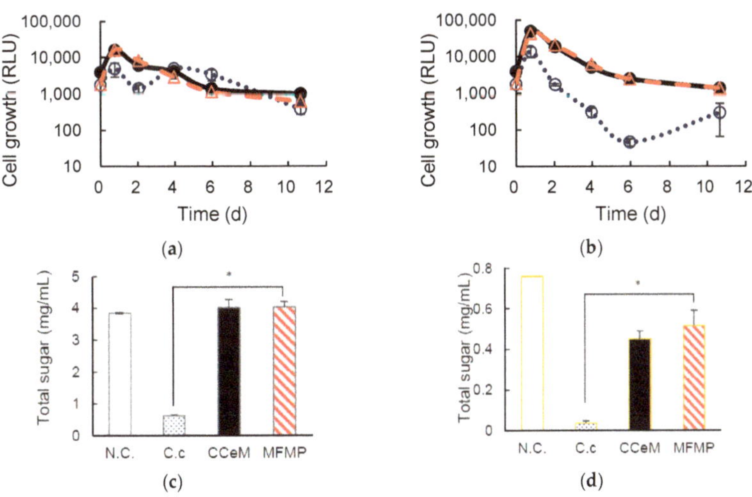

Figure 3. *Cont.*

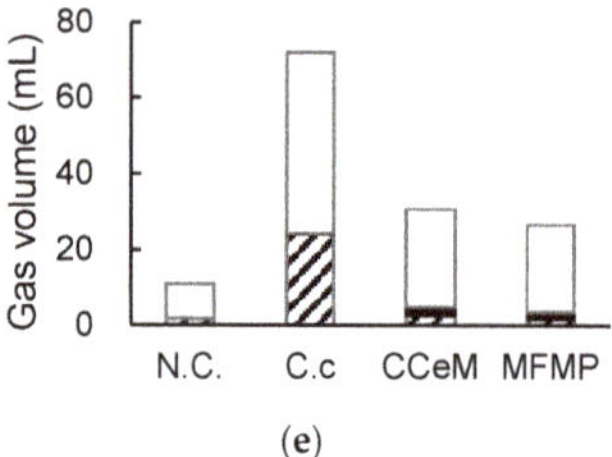

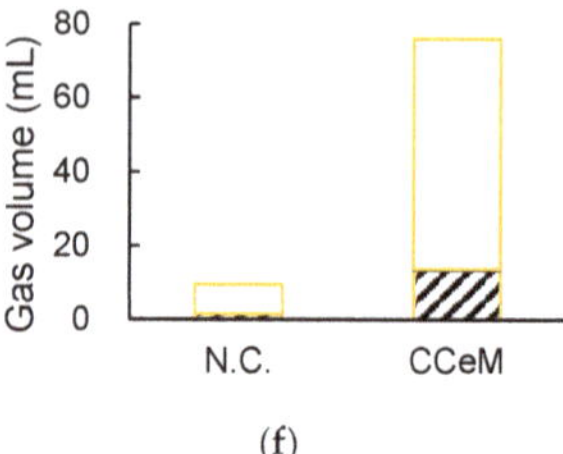

(e)

(f)

Figure 3. Cultivation of *C. cellulovorans*, cellobiose and MFMP (CCeM) and microbial flora of methane production (MFMP) with Avicel. (**a**) Cell growth, where *C. cellulovorans* (○_open circle), CCeM (●_closed circle) and MFMP (Δ_open triangle) are included; (**c**) total sugar concentration; (**e**) gas production of H_2 (hatched bar), CH_4 (closed bar), and CO_2 (open bar) are included. Cultivation of *C. cellulovorans*, CCeM, and MFMP with removed peel. (**b**) Cell growth, where *C. cellulovorans* (○_open circle), CCeM (●_closed circle), and MFMP (Δ_open triangle) are included; (**d**) total sugar concentration; (**f**) gas production of H_2 (hatched bar), CH_4 (closed bar), and CO_2 (open bar) are included. Values with error bars are mean ± SE of three independent samples. SE means a standard error. An asterisk indicates a significant difference ($p < 0.05$).

3.2. Cellulose Degradation and Methane Production under Cocultivation with Methanogens and the Armed C. cellulovorans

C. cellulovorans was pre-cultivated in media containing 0.5% (*w/v*) of MN301 for 2 days, which meant *C. cellulovorans* was armed with cellulosome. Anaerobic batch cultivations of the armed *C. cellulovorans*, ACCeM, and MFMP were carried out in a 40-mL medium containing 0.5% (*w/v*) of MN301 at 37 °C without shaking. RLU values of the preculture of the armed *C. cellulovorans* and MFMP were 6786 and 38,538, respectively. RLU value of the armed *C. cellulovorans* culture did not increase over 10,000 like the non-armed *C. cellulovorans*. Therefore, inoculation amounts for monocultures of the armed *C. cellulovorans* and MFMP were decided so that initial RLU values of each monoculture became close to 1000. Inoculation volumes of the armed *C. cellulovorans* and MFMP were decided as 7 mL and 1.2 mL, respectively. The armed *C. cellulovorans* was 6.69 times dilution and MFMP was 40.17 times dilution so that initial RLU values of monoculture of the armed *C. cellulovorans* and MFMP were 1014 and 959, respectively. ACCeM was inoculated in 7 mL of the armed *C. cellulovorans* preculture and 1.2 mL of MFMP preculture so that concentrations of cell growth against substrate became the same as the monoculture. Cell growths in the armed *C. cellulovorans* monoculture, ACCeM culture, and MFMP monoculture were measured for 7 days of cultivation. RLU values in ACCeM culture and MFMP monoculture rapidly increased more than 100,000 for 10 h of cultivation. The RLU value in the armed *C. cellulovorans* monoculture increased for 2 days of cultivation and then decreased slowly. Interestingly, the RLU value in ACCeM culture was higher than that in MFMP monoculture after 2 days of cultivation. It suggested that the armed *C. cellulovorans* grew for a couple of days of cultivation and coexisted with MFMP (Figure 4a). The total sugar concentration in the armed *C. cellulovorans* monoculture decreased to 0.38 mg/mL from 5 mg/mL for 7 days of cultivations; it indicated that 92.4% of MN301 was degraded by the armed *C. cellulovorans* for 1 week (Figure 4b). The total sugar concentration in ACCeM culture did not decrease for 3 days of cultivation, however it rapidly decreased from 4 days of cultivation. It also suggested that the armed *C. cellulovorans* survived for 3 days using brought in cellulosome while adapting to coexisting MFMP. Total sugar concentration in ACCeM culture decreased to 1.08 mg/mL for 7 days of cultivation, therefore it was demonstrated that ACCeM could degrade 78.4% of MN301 for 1 week. The total sugar concentration in MFMP monoculture did not decrease for 27 days of cultivation, it indicated that MFMP did not have a capability to degrade MN301. The total sugar concentrations in the armed *C. cellulovorans* monoculture and ACCeM culture was significantly low than that in MFMP for 7 days of cultivation (Figure 4c). Total gas volume in the ACCeM culture rapidly increased for 10 h of cultivation, and the total gas volume kept for 1–7 days of cultivation. Interestingly, the methane proportion in the total gas increased for 1–7 days of cultivation, even though total gas volume did not increase. Furthermore, methane volume continuously increased for 21 days of cultivation. Finally, methane volume increased 6.89 mL for 1–21 days of cultivation, therefore it revealed that methane production rate from MN301 was 0.014 mL/h (Figure 4d). Organic acid concentrations in the culture supernatant of ACCeM were measured. Lactic acid concentration was slightly detected for 1 day of cultivation and was not detected after 2 days of cultivation (Figure 4e). Formic acid concentration was temporally increased at 1 day of cultivation, and then it was not detected after 2 days of cultivation. Acetic acid concentration increased for 6 days of cultivation that was the same term when the total sugar concentration reached 78.4%. Acetic acid concentration turned to decrease from 8 days of cultivation. On the other hand, propionic acid concentration increased but did not turn to decrease. It was suggested that MN301 was degraded by *C. cellulovorans* and acetic acid and methane were produced, and methane was continuously produced utilizing acetic acid after cellulose was degraded. Moreover, it suggested that acetic acid was converted to methane in preference to propionic acid.

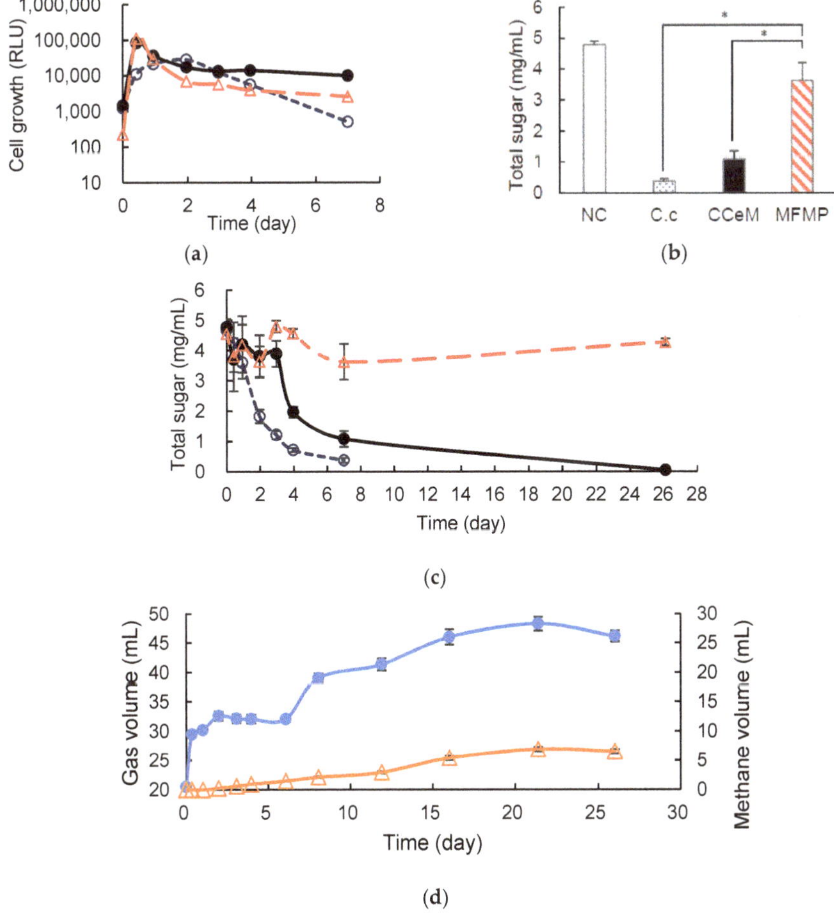

Figure 4. *Cont.*

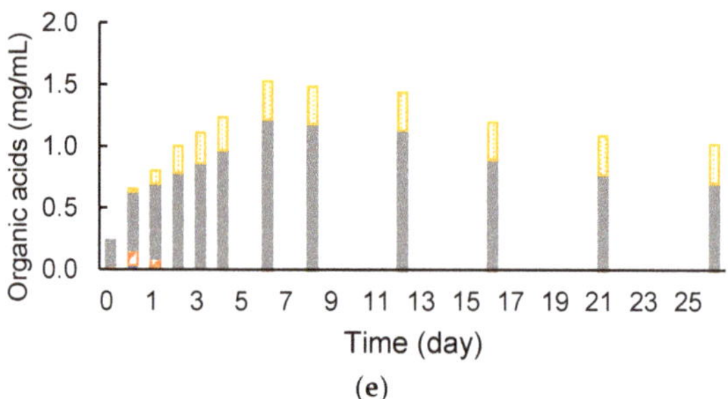

(e)

Figure 4. Cultivation of *C. cellulovorans*, CCeM, and MFMP with Avicel. (**a**) Cell growth, where *C. cellulovorans* (○_open circle), CCeM (●_closed circle), and MFMP (Δ_open triangle) are included; (**b**) total sugar concentration; (**c**) time course of total sugar, where *C. cellulovorans* (○_open circle), CCeM (●_closed circle), and MFMP (Δ_open triangle) are included; (**d**) time course of gas production, where total gas volume (●_closed circle) and methane volume (Δ_open triangle) are included; (**e**) time course of organic acids concentration of CCeM, where lactic acid (closed bars), formic acid (open bars), acetic acid (closed bars), and propionic acid (dotted bars) are included. Values with error bars are mean ± SE of three independent samples. SE means a standard error. An asterisk indicates a significant difference ($p < 0.05$).

3.3. Removed Peel Degradation and Methane Production under Cocultivation with Methanogens and the Armed C. cellulovorans

It was confirmed that the armed *C. cellulovorans* could degrade cellulose and MFMP also worked to produce methane simultaneously. Anaerobic batch cultivations of ACCeM was carried out in a 20-mL medium containing 0.5% (*w/v*) of removed peel, which was an actual agricultural waste, at 37 °C without shaking. The armed *C. cellulovorans* was pre-cultivated in media containing 0.5% (*w/v*) of MN301 for 2 days. According to measured cell growth of precultures, inoculation amounts for monocultures of the armed *C. cellulovorans* and MFMP were decided so that initial RLU values of each monoculture closely became 1000. RLU values of the preculture of the armed *C. cellulovorans* and MFMP were 3448 and 38,538, respectively. Inoculation volumes of the armed *C. cellulovorans* and MFMP were decided at 8.5 mL and 0.8 mL, respectively. The armed *C. cellulovorans* was 3.44 times dilution and MFMP was 36.6 times dilution so that the initial RLU values of ACCeM were 1002 for the armed *C. cellulovorans* and 1052 for MFMP. The degradation of removed peel was observed for 5 days of cultivation (Figure 5a). Gas production started immediately after inoculation and continued for 4 days of cultivation. However, methane was not detected for these early 4 days of cultivation (Figure 5b). Methane began to be detected after 12 days of cultivation. Interestingly, the increment of methane was 4.7 mL for 12–26 days of cultivation against that the total gas was 7.1 mL for the same term, therefore methane occupied 66.2% of the increased gas. This methane concentration is good performance as the fuel for a biogas power generation. Furthermore, methane productivity was 0.014 mL/h for 12–26 days of cultivation, which was equal to that of MN301 substrate. Acetic acid concentration in the culture supernatant increased the same amount as the MN301 substrate. However, the acetic acid concentration was 2.5 times higher than that in the culture supernatant with MN301 (Figure 5c). It suggested that there was room to convert much acetic acid to methane to improve methane production.

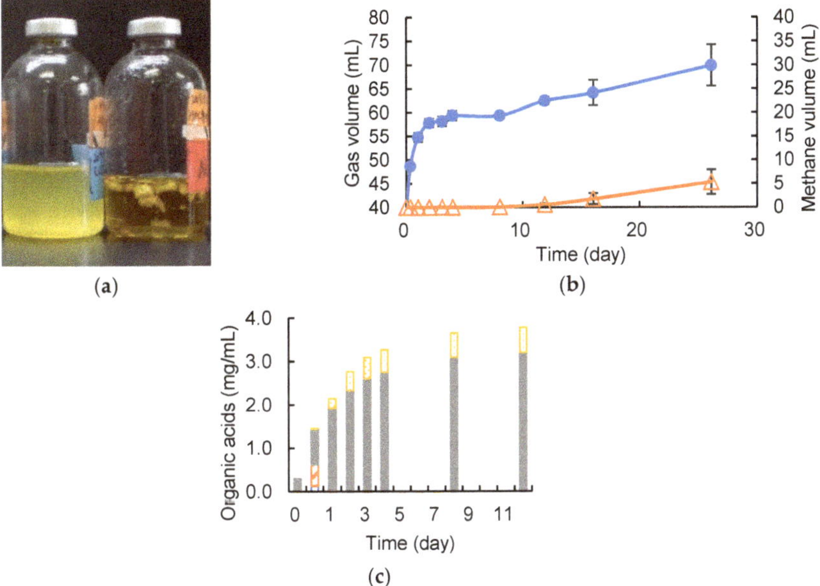

Figure 5. (a) CCeM culture (left), negative control (right); (b) time course of gas production, where total gas volume (●_closed circle) and methane volume (Δ_open triangle) are included; (c) time course of organic acids concentration of CCeM, where lactic acid (closed bars), formic acid (open bars), acetic acid (closed bars), and propionic acid (dotted bars) are included. Values with error bars are mean ± SE of three independent samples. SE means a standard error. An asterisk indicates a significant difference ($p < 0.05$).

4. Discussion

Since reducing carbon dioxide to conserve the global environment is one of the purposes to replace fossil fuels with biofuels, it is desirable to adopt the biofuel process that has a low environmental load. A sulfuric acid degradation of cellulosic biomass brings about corrosion of a reactor and needs a waste liquid treatment, and the steam explosion process requires a lot of energy to create high pressure and temperature [25,26], therefore these are not low environmental load methods. The enzymatic saccharification is environmental-friendly, because it is a mesophilic process and does not use acids. However, the degradation cost is high because the purified enzymes are expensive and used in large quantities. Instead of using expensive enzymes, the cost will be reduced to utilize microorganisms that can produce enzymes. Furthermore, since *C. cellulovorans* has many genomes related to cellulosome and non-cellulosomal [27], there is potential to improve the efficiency of the enzymatic saccharification. In addition, when the sulfuric acid process or the steam explosion process is used to saccharide cellulosic biomass, various microorganisms attached to the biomass and brought into the saccharification process are not a problem, because they are treated by high temperature and heat or acid. However, there is the key issue that the microorganisms which produce enzymes must survive and perform its saccharification ability coexisting other various microorganisms, because a lot of microorganisms are brought into the culture apparatus together with cellulosic biomass. In this respect, the present study provided a method that enables the co-culture of various microbiota and *C. cellulovorans*, which has been difficult until now. Although there are few reports only co-culture *C. cellulovorans* and one other microbe [28], many experiences can be performed from cellulose to methane using a consortium and the armed *C. cellulovorans*. Notably, this armed *C. cellulovorans* is non-GMO (non- Genetically Modified Organisms) and is not modified with genome editing, such as CRISPR/Cas9 [29], therefore anyone can easily cultivate the armed *C. cellulovorans* from a cellulosic substrate medium and use it anywhere. It is useful and powerful that the armed *C. cellulovorans*, which anyone can use, degrades orange wastes including d-limonene and the production of methane, which accounts for 66% of the total gas volume, was produced from the consortia.

However, for stable biogas production, it is essential that the carbon, nitrogen, phosphorus, sulfur, and a carbon-to-nitrogen ratio (C/N ratio) in the culture medium have the desired values. In order for sustainable biogas production from cellulosic biomass, it is necessary to investigate in detail how these components are indispensable for microbial growth change. On the other hand, since the sulfur content contained in the cellulosic biomass is small, simplification of a desulphurization equipment in the biogas plant can be expected.

Moreover, an orange contains beta-cryptoxanthin (β-cry), especially in the orange peel [30]. β-cry is a natural carotenoid pigment, and carotenoids are known to contribute to the defense system of the human body against reactive oxygen species. Inverse associations of serum β-cry with the risk for cancer, diabetes, and liver dysfunction have been reported [31–34]. The enzymatic degradation of orange peel can extract carotenoids contained in the culture. This means orange peel, which was disposed of so far, can be converted to useful raw material to have functional foods. In addition, the enzymatic saccharification does not decompose the compound by high temperature and pressure and by acid, therefore the enzymatic saccharification is the most advanced process to extract the useful compounds.

5. Conclusions

This study revealed that *C. cellulovorans* by growing with cellulose instead of cellobiose can coexist with complex microbiota such as methanogenic microbiota. Therefore, it provided a useful method for researching the co-culture of *C. cellulovorans* with various other microorganisms. Furthermore, *C. cellulovorans* and methanogenic microflora coexisted with each other while keeping the cellulose degradation ability and the methane production ability. As a result, the biogas was produced containing 66.2% of methane using mandarin orange peel as a carbon source, and when the biogas contains 60% or more of methane, the biogas can be used directly as fuel for a gas engine.

Author Contributions: Conceptualization, Y.T. and H.T.; methodology, Y.T. and H.T.; validation, Y.T. and H.T.; resources, Y.T.; data curation, H.T.; writing—original draft preparation, H.T.; writing—review and editing, Y.T.; supervision, Y.T.; project administration, H.T.; funding acquisition, Y.T.

Funding: This research received no external funding.

Acknowledgments: We would like to thank Daimasa Engineering Co. LTD for sample preparation and assignment.

Conflicts of Interest: The authors declare no conflict of interest.

References

1. Naik, S.N.; Goud, V.V.; Rout, P.K.; Dalai, A.K. Production of first and second generation biofuels: A comprehensive review. *Renew. Sustain. Energy Rev.* **2010**, *14*, 578–597. [CrossRef]
2. Schenk, P.M.; Thomas-Hall, S.R.; Stephens, E.; Marx, U.C.; Mussgnug, J.H.; Posten, C.; Kruse, O.; Hankamer, B. Second generation biofuels: High-efficiency microalgae for biodiesel production. *Bioenergy Res.* **2008**, *1*, 20–43. [CrossRef]
3. Brethauer, S.; Studer, M.H. Biochemical conversion processes of lignocellulosic biomass to fuels and chemicals—A review. *Chimia* **2015**, *69*, 572–581. [CrossRef] [PubMed]
4. Gray, K.A.; Zhao, L.; Emptage, M. Bioethanol. *Curr. Opin. Chem. Biol.* **2006**, *10*, 141–146. [CrossRef]
5. Grohmann, K.; Baldwin, E.A.; Buslig, B.S. Production of ethanol from enzymatically hydrolyzed orange peel by the yeast Saccharomyces cerevisiae. *Appl. Environ. Microbiol.* **1994**, *45*, 315–327.
6. Winniczuk, P.P.; Parish, M.E. Minimum inhibitory concentrations of antimicrobials against microorganisms related to citrus juice. *Food Microbiol.* **1997**, *14*, 373–381. [CrossRef]
7. Lane, A.G. Removal of peel oil from citrus peel press liquors before anaerobic digestion. *Environ. Technol. Lett.* **1983**, *4*, 65–72. [CrossRef]

8. Pourbafrani, M.; Talebnia, F.; Niklasson, C.; Taherzade, M.J. Protective effect ofencapsulation in fermentation of limonene-contained media and orange peel hydrolyzate. *Int. J. Mol. Sci.* **2007**, *8*, 777–787. [CrossRef]

9. Tomita, H.; Tamaru, Y. Direct IBE fermentation from mandarin orange wastes by combination of *Clostridium cellulovorans* and *Clostridium beijerinckii*. *AMB Express* **2019**, *9*, 1. [CrossRef]

10. Jones, D.T.; Woods, D.R. Acetone-Butanol Fermentation Revisited. *Microbiol. Rev.* **1986**, *50*, 484–524.

11. Vieiraa, C.F.S.; Filhob, F.M.; Filhoa, R.M.; Marianoa, A.P. Acetone-free biobutanol production: Past and recent advances in the Isopropanol-Butanol-Ethanol (IBE) fermentation. *Bioresour. Technol.* **2019**, *287*, 121425. [CrossRef] [PubMed]

12. Doi, R.H.; Kosugi, A. Cellulosomes: Plant-cell-wall-degrading enzyme complexes. *Nat. Rev. Microbiol.* **2004**, *2*, 541–551. [CrossRef] [PubMed]

13. Sleat, R.; Mah, R.A.; Robinson, R. Isolation and characterization of an anaerobic, cellulolytic bacterium, *Clostridium cellulovorans* sp. nov. *Appl. Environ. Microbiol.* **1984**, *48*, 88–93. [PubMed]

14. Koukiekolo, R.; Cho, H.Y.; Kosugi, A.; Inui, M.; Yukawa, H.; Doi, R.H. Degradation of corn fiber by *Clostridium cellulovorans* cellulases and hemicellulases and contribution of scaffolding protein CbpA. *Appl. Environ. Microbiol.* **2005**, *71*, 3504–3511. [CrossRef]

15. Beukes, N.; Chan, H.; Doi, R.H.; Pletschke, B.I. Synergistic associations between *Clostridium cellulovorans* enzymes XynA, ManA and EngE against sugarcane bagasse. *Enzym. Microb. Technol.* **2008**, *42*, 492–498. [CrossRef]

16. Dredge, R.; Radloff, S.E.; van Dyk, J.S.; Pletschke, B.I. Lime pretreatment of sugar beet pulp and evaluation of synergy between ArfA, ManA and XynA from *Clostridium cellulovorans* on the pretreated substrate. *3 Biotech* **2011**, *1*, 151–159. [CrossRef]

17. Tamaru, Y.; Miyake, H.; Kuroda, K.; Ueda, M.; Doi, R.H. Comparative genomics of the mesophilic cellulosome-producing *Clostridium cellulovorans* and its application to biofuel production via consolidated bioprocessing. *Environ. Technol.* **2010**, *31*, 889–903. [CrossRef]

18. Matsui, K.; Bae, J.; Esaka, K.; Morisaka, H.; Kuroda, K.; Ueda, M. Exoproteome profiles of *Clostridium cellulovorans* on various carbon sources. *Appl. Environ. Microbiol.* **2013**, *79*, 6576–6584. [CrossRef]

19. Tamaru, Y.; Ui, S.; Murashima, K.; Kosugi, A.; Chan, H.; Doi, R.H.; Liu, B. Formation of protoplasts from cultured tobacco cells and Arabidopsis thaliana by the action of cellulosomes and pectate lyase from *Clostridium cellulovorans*. *Appl. Environ. Microbiol.* **2002**, *68*, 2614–2618. [CrossRef]

20. Blair, B.G.; Anderson, K.L. Comparison of staining techniques for scanning electron microscopic detection of ultrastructural protuberances on cellulolytic bacteria. *Biotech. Histochem.* **1998**, *73*, 107–113. [CrossRef]

21. Blair, B.G.; Anderson, K.L. Regulation of cellulose inducible structures of *Clostridium cellulovorans*. *Can. J. Microbiol.* **1999**, *45*, 242–249. [CrossRef] [PubMed]

22. Guo, J.; Peng, Y.; Ni, B.J.; Han, X.; Fan, L.; Yuan, Z. Dissecting microbial community structure and methane-producing pathways of a full-scale anaerobic reactor digesting activated sludge from wastewater treatment by metagenomic sequencing. *Microb. Cell Factories* **2015**, *14*, 33. [CrossRef] [PubMed]

23. Tomita, H.; Tamaru, Y. Biomethane production from sugar beet pulp under cocultivation with *Clostridium cellulovorans* and methanogens. *AMB Express* **2019**, *9*, 28. [CrossRef] [PubMed]

24. Miyake, H.; Maeda, Y.; Ishikawa, T.; Tanaka, A. Calorimetric studies of the growth of anaerobic microbes. *J. Biosci. Bioeng.* **2016**, *122*, 364–369. [CrossRef]

25. Joshi, S.M.; Waghmare, J.S.; Sonawane, K.D.; Waghmare, S.R. Bio-ethanol and bio-butanol production from orange peel waste. *Biofuels* **2015**, *6*, 55–56. [CrossRef]

26. Oberoi, H.S.; Vadlani, P.V.; Madl, R.L.; Saida, L.; Abeykoon, J.P. Ethanol Production from Orange Peels: Two-Stage Hydrolysis and Fermentation Studies Using Optimized Parameters through Experimental Design. *J. Agric. Food Chem.* **2010**, *58*, 3422–3429. [CrossRef]

27. Tamaru, Y.; Miyake, H.; Kuroda, K.; Nakanishi, A.; Matsushima, C.; Doi, R.H.; Ueda, M. Comparison of the mesophilic cellulosome-producing *Clostridium cellulovorans* genome with other cellulosome-related clostridial genomes. *Microb. Biotechnol.* **2011**, *4*, 64–73. [CrossRef]

28. Lu, H.; Ng, S.K.; Jia, Y.; Cai, M.; Lee, P.K.H. Physiological and molecular characterizations of the interactions in two cellulose-to-methane cocultures. *Biotechnol. Biofuels* **2017**, *10*, 37. [CrossRef]

29. Jinek, M.; Chylinski, K.; Fonfara, I.; Hauer, M.; Doudna, J.A.; Charpentier, E. A Programmable Dual-RNA–Guided DNA Endonuclease in Adaptive Bacterial Immunity. *Science* **2012**, *337*, 816–821. [CrossRef]

30. Holden, J.M.; Eldridge, A.L.; Beecher, G.R.; Buzzard, I.M.; Bhagwat, S.; Davis, C.S.; Douglas, L.W.; Gebhardt, S.; Haytowitz, D.; Schakel, S. Carotenoid content of U.S. foods: An update of the database. *J. Food Compos. Anal.* **1999**, *12*, 169–196. [CrossRef]

31. Sluijs, I.; Beulens, J.W.; Grobbee, D.E.; van der Schouw, Y.T. Dietary carotenoid intake is associated with lower prevalence of metabolic syndrome in middle-aged and elderly men. *J. Nutr.* **2009**, *139*, 987–992. [CrossRef] [PubMed]

32. Lorenzo, Y.; Azqueta, A.; Luna, L.; Bonilla, F.; Domínguez, G.; Collins, A.R. The carotenoid β-cryptoxanthin stimulates the repair of DNA oxidation damage in addition to acting as an antioxidant in human cells. *Carcinogenesis* **2009**, *30*, 308–314. [CrossRef] [PubMed]

33. Burri, B.J. Beta-cryptoxanthin as a source of vitamin A. *J. Sci. Food Agric.* **2014**, *95*, 1786–1794. [CrossRef] [PubMed]

34. Lian, F.; Hu, K.Q.; Russell, R.M.; Wang, X.D. β-Cryptoxanthin suppresses the growth of immortalized human bronchial epithelial cells and non-small-cell lung cancer cells and up-regulates retinoic acid receptor β expression. *Cancer Cell Biol.* **2006**, *119*, 2084–2089. [CrossRef] [PubMed]

 *fermentation*

Article

Developing a Microbial Consortium for Enhanced Metabolite Production from Simulated Food Waste

Nathan D. Schwalm III [1,*], Wais Mojadedi [2], Elliot S. Gerlach [1], Marcus Benyamin [2], Matthew A. Perisin [1] and Katherine L. Akingbade [1,*]

[1] Combat Capabilities Development Command Army Research Laboratory, FCDD-RLS-EB, 2800 Powder Mill Road, Adelphi, MD 20783, USA; elliot.s.gerlach.civ@mail.mil (E.S.G.); matthew.a.perisin.civ@mail.mil (M.A.P.)

[2] Oak Ridge Associated Universities, 4692 Millennium Drive, Suite 101, Belcamp, MD 21017, USA; wais.mojadedi.ctr@mail.mil (W.M.); marcus.s.benyamin2.ctr@mail.mil (M.B.)

* Correspondence: nathan.d.schwalm.civ@mail.mil (N.D.S.III); katherine.l.akingbade.civ@mail.mil (K.L.A.); Tel.: +301-394-0193 (N.D.S.III); +301-394-0604 (K.L.A.)

Received: 25 September 2019; Accepted: 22 November 2019; Published: 27 November 2019

Abstract: Food waste disposal and transportation of commodity chemicals to the point-of-need are substantial challenges in military environments. Here, we propose addressing these challenges via the design of a microbial consortium for the fermentation of food waste to hydrogen. First, we simulated the exchange metabolic fluxes of monocultures and pairwise co-cultures using genome-scale metabolic models on a food waste proxy. We identified that one of the top hydrogen producing co-cultures comprised *Clostridium beijerinckii* NCIMB 8052 and *Yokenella regensburgei* ATCC 43003. A consortium of these two strains produced a similar amount of hydrogen gas and increased butyrate compared to the *C. beijerinckii* monoculture, when grown on an artificial garbage slurry. Increased butyrate production in the consortium can be attributed to cross-feeding of lactate produced by *Y. regensburgei*. Moreover, exogenous lactate promotes the growth of *C. beijerinckii* with or without a limited amount of glucose. Increasing the scale of the consortium fermentation proved challenging, as two distinct attempts to scale-up the enhanced butyrate production resulted in different metabolic profiles than observed in smaller scale fermentations. Though the genome-scale metabolic model simulations provided a useful starting point for the design of microbial consortia to generate value-added products from waste materials, further model refinements based on experimental results are required for more robust predictions.

Keywords: butyrate; *Clostridium beijerinckii*; cross-feeding; food waste; genome-scale metabolic model; hydrogen; lactate; *Yokenella regensburgei*

1. Introduction

Generation of food waste is a global problem [1,2]. Approximately 33% of food produced is wasted, contributing to 30–60% of the total solid waste produced worldwide [3]. The burden of food waste disposal extends to military environments, where it imposes an additional logistical burden on Army operations [4]. The primary waste produced by the soldier is food related trash [4], and current practices dictate that field generated waste must either be buried or burned daily [5]. An alternative to these practices is the anaerobic fermentation of food waste to commodity chemicals [1]. Anaerobic fermentation of a single carbon feedstock is frequently used in industrial-scale processes to produce biofuels [6] and commodity chemicals [7]. Significant challenges exist applying anaerobic fermentations to food waste breakdown, due to the complexity and inconsistency of the fermentation feedstock. Utilizing additional microbes that can support the primary fermentation strain as part of a microbial consortium is a potentially promising avenue to increasing the viability of food waste to value-added chemical conversion.

Considerable research efforts are underway to harness both microbial consortia comprising multiple genotypes of the same species [8,9] and multi-species consortia [10,11] for improved production of chemicals from readily available complex substrates [12,13]. Microbial consortia enable separation of complex substrate breakdown and chemical production between different species, which can reduce the metabolic burden on individual cells and reduce the need to engineer these microbes [14,15]. Cross-feeding of metabolites from extracellular substrate breakdown [16], cellular secretion [17], or via direct cell-to-cell interactions [18,19] can support the stability of a microbial consortium. These interactions can be designed to force the dependence of auxotrophic strains on each other for growth [20] or can be a result of native metabolic byproducts from one organism acting as a substrate for another organism [21].

Hydrogen is a promising alternative fuel source to petrochemicals due to its high energy content (120 MJ/kg versus 46.7 MJ/kg for gasoline) [22]. Microbes primarily produce hydrogen via photofermentation by the purple non-sulfur bacteria *Rhodobacter* and *Rhodopseudomonas* [23], and during dark fermentation by strictly anaerobic *Clostridium* species [22]. In purple non-sulfur bacteria, hydrogen production is coupled with organic acid utilization [23]. In *Clostridium* species, hydrogen production is coupled with ferredoxin oxidation and is accompanied by organic acid production [24], which inhibits hydrogenase activity and hydrogen production [25]. Microbial consortia of *Clostridium* that generate organic acids and purple non-sulfur bacteria that use organic acids for hydrogen production are being explored [26,27], as are consortia comprising multiple *Clostridium* species [13,28]. The formate-hydrogen lyase complex, found mainly in *Proteobacteria*, produces hydrogen via the anaerobic breakdown of formate [29]. This complex is also being explored as a route for hydrogen production in *Escherichia coli* engineered to express a hydrogenase [29], and as part of microbial consortia with *Clostridium* [30].

Genome scale metabolic models (GSMMs), combined with flux-balance analysis (FBA), provide predictive frameworks not only to assess the metabolic outputs from a variable inputs, but also to assess metabolic interactions between multiple microbes. These models are built from enzymatic genome annotations that determine available reactions for each microbe, resulting in a table of reactants/products by reaction [31]. These tables are then fed to FBA to determine what metabolic outputs are possible given a set of inputs, assuming a steady-state [32]. To further limit the solution space, there is an assumption that each microbe will optimize growth (i.e., biomass production) [32]. This framework has been demonstrated to accurately recapitulate single microbe metabolisms and metabolic interactions between members of a consortium [33–37].

Recently, Magnusdottir et al. created GSMMs of approximately equal refinement for 773 gut microbes and simulated all monocultures and co-cultures on Western and high-fiber diet inputs [38]. Ecological interactions between these microbes were predicted to vary based on the diet and whether the environment was aerobic or anaerobic [38]. Perisin and Sund followed up on these findings to predict whether combinations of microbes could produce specialty/commodity chemicals at higher rates than monocultures [39]. Co-cultures were identified that could overproduce chemicals of interest such as hydrogen gas, butanol, methane, formaldehyde, propionate, and urea [39].

We developed a framework to experimentally test commodity chemical production from microbial consortia predicted to synergistically overproduce an individual chemical (Figure 1). We used genome-scale metabolic models and flux balance analysis of human gut microbiota strains provided a simulated food waste to identify two species bacterial consortia with the greatest synergistic production of hydrogen gas. We then measured the gases and metabolites produced by one of the highest overproducing consortia and its component species when cultured on an artificial garbage slurry medium (AGS). Moreover, we demonstrated the cross-feeding of metabolites between the two species. Finally, we attempted to scale-up the AGS fermentation with two distinct conditions, stationary fermentation or stirred fermentation with pH control, to assess the feasibility of the method. Experimental consortium fermentations will further improve metabolic models for future experimental testing.

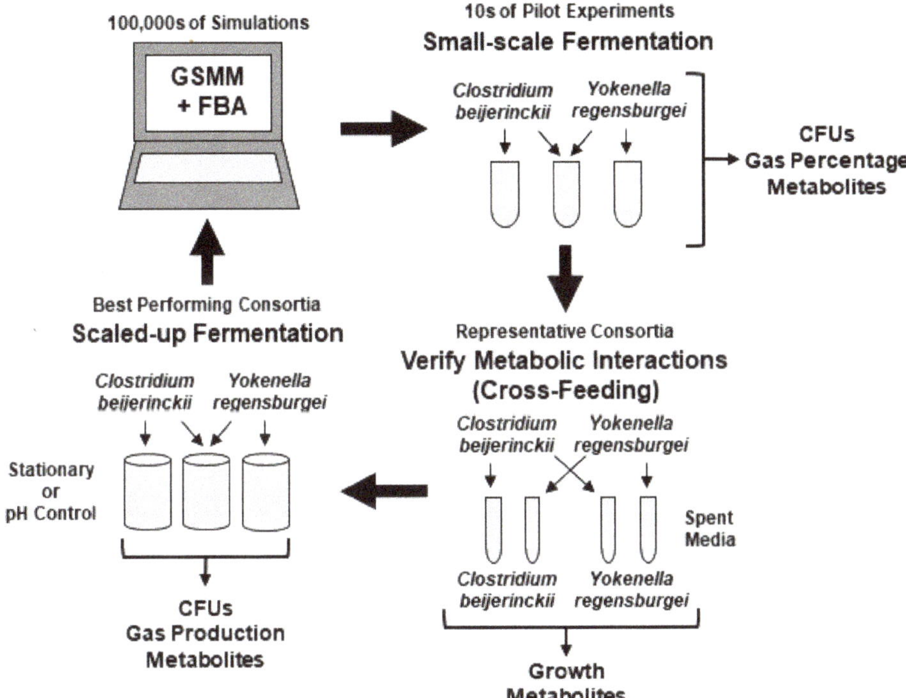

Figure 1. Experimental framework. Genome-scale metabolic modeling (GSMM) and flux balance analysis (FBA) were used to predict the metabolic outputs of 298,378 pairwise combinations of microbial consortia provided a simulated food waste medium. The consortium of *Clostridium beijerinckii* and *Yokenella regensburgei* was predicted to have the second highest synergistic overproduction of hydrogen compared to any individual strain. Small-scale fermentations were performed using the *C. beijerinckii* and *Y. regensburgei* consortium on an artificial garbage slurry medium to simulate food waste. Growth (CFUs), metabolites, and percentage of gases produced in gas chromatography vials were measured. Spent media from cultures of *C. beijerinckii* and *Y. regensburgei* were used to cross-feed to the other species, and growth and metabolite production were measured. Small-scale experiments were used to inform design of scaled-up fermentations of monocultures and the consortium of *C. beijerinckii* and *Y. regensburgei*. The results of these experiments can be used to further improve the GSMM and FBA forming a broad framework for testing production of commodity chemicals by microbial consortia.

2. Materials and Methods

2.1. Genome-Scale Metabolic Modeling

Simulations were performed as in Perisin and Sund [39]. Genome-scale metabolic models (GSMMs) for *Clostridium beijerinckii* NCIMB 8052 and *Yokenella regensburgei* ATCC 43003 (AGORA v 1.01) were downloaded from the Virtual Metabolic Human database (http://vmh.uni.lu) [38]. Lower bounds for input exchange reactions were set to mimic Western diet food waste (Supplementary Table 1 in [39]). All simulations were performed in R v3.4.0 [40] with plots created using *ggplot2* v 2.2.1 [41]. Systems Biology Markup Language (SBML) models were first uploaded with the *sybilSBML* v 3.0.1 [42] package. Then, *sybil* v2.0.4 [43], and *glpkAPI* v 1.3.0 [44] packages were used for model manipulations and flux-balance analysis (FBA). For monoculture simulations, flux through each model's biomass reaction was maximized. The maximum biomass flux was then used to run parsimonious enzyme usage FBA (pFBA, *mtf* algorithm in *sybil*) so that the total absolute flux was minimized. For co-culture simulations, models were combined in a similar manner to Magnusdottir et al. [38]. The COBRA [45] MATLAB script, createMultipleSpeciesModel.m found at: https://github.com/opencobra/cobratoolbox, which is based on the FBA implementation in Klitgord and Segre [46], was used as a template. This script was converted to work in R with the *sybil* package and creates a common environment for metabolites to be exchanged between models, but does not create a host compartment as in the COBRA implementation. After combining the models, input exchange fluxes were updated based on Western diet food waste as above, and pFBA was used to simulate growth, simultaneously maximizing each model's growth and minimizing the total absolute flux.

2.2. Bacterial Strains and Growth Conditions

Clostridium beijerinckii NCIMB 8052 (ATCC 51743) and *Yokenella regensburgei* ATCC 43,003 were obtained from ATCC. Both strains were cultivated in anaerobic conditions (5% carbon dioxide, 5% hydrogen, 90% nitrogen) at 37 °C in a Coy anaerobic chamber and maintained as stocks at −80 °C in media containing 20% glycerol. *C. beijerinckii* was routinely cultured on Clostridial Growth Medium (CGM) [47] or Luria Bertani (LB) medium (Fisher, Fair Lawn, NJ, USA), supplemented with 0.5% glucose, unless otherwise noted. Thiamphenicol (30 µg mL^{-1}) was used to select for *C. beijerinckii* when necessary [48,49]. *Y. regensburgei* was routinely cultured on LB supplemented with 0.5% glucose or Brain Heart Infusion (BHI) medium (Oxoid, Lenexa, KS, USA). Erythromycin (40 µg mL^{-1}) and aerobic growth at 37 °C were used to select for *Y. regensburgei*, when necessary [49]. The artificial garbage slurry was generated to model organic solid waste, as previously described [50], with limited modifications. Briefly, 10% (w v^{-1}) dog food (Kibble 'n Bits® Chef Choice Bistro Oven Roasted Beef, Spring Vegetable & Apple Flavor) was blended in distilled water and autoclaved (121 °C, 30 min exposure). Following the first autoclave cycle, the pH was balanced to 7.0 via addition of sodium hydroxide solution and buffered to 50 mM potassium phosphate pH 6.7, prior to a second autoclave cycle. For cross-feeding experiments, *C. beijerinckii* or *Y. regensburgei* were grown in LB supplemented with 0.5% glucose for 24 h and filtered through a 0.2 µm PES membrane (Corning, Corning, NY, USA) to produce spent media. Spent media with the addition of 50 mM potassium phosphate pH 6.8 was used for an additional 24 h of culturing of either *C. beijerinckii* or *Y. regensburgei*. For experiments assessing the ability of *C. beijerinckii* to utilize exogenous lactate, the indicated amount of DL-lactic acid (Fluka, St. Louis, MO, USA) was added to cultures and neutralized with sodium hydroxide. Bacterial growth was assessed by measuring absorbance at 600 nm, for cross-feeding and lactate supplementation experiments, or plating of multiple dilutions to count colony forming units per milliliter (CFU mL^{-1}) on selective media, for experiments using AGS.

2.3. Bioreactor Fermentations

Bioreactor fermentations were conducted in a DASGIP® parallel fermentation system (Eppendorf, Jülich, Germany). AGS was brought to pH 6.7 by addition of sodium hydroxide solution, following the addition of the 50 mM potassium phosphate and aliquoted to reaction vessels prior to autoclaving (121 °C, 30 min exposure) a single time. For the pH-controlled bioreactor experiment, the DASGIP® system automatically added sodium hydroxide to maintain a pH of at least 6.7. Temperature was maintained at 37 °C and anaerobic headspace maintained via nitrogen addition at a flow rate of 1 L min^{-1}.

2.4. HPLC Quantification of Metabolites

Fermentation products were separated and quantified using an Aminex HPX-87H organic acid column (300 × 7.8 mm; Bio-Rad, Hercules, CA, USA) on an Agilent 1200 series HPLC (Agilent Technologies, Santa Clara, CA, USA) equipped with a multi-wavelength UV/vis detector (210 nm and 280 nm) and refractive index detector. Samples were filtered through a 0.2 μm PES membrane (Corning) and stored at 4 °C until applied to the column with an injection volume of 20 μL. Separation was performed using an isocratic mobile phase consisting of 3.25 mM sulfuric acid at a flow rate of 0.600 mL min^{-1} for a total run time of 55 min, and was temperature controlled for the entire duration at 35 °C. Quantification of acetate, acetone, butyrate, ethanol, formate, glucose, and lactate was performed through the use of standardized concentration gradients with HPLC-grade standards obtained from various vendors (not shown). Data was processed using ChemStation (Agilent).

2.5. Gas Chromatography

Hydrogen and carbon dioxide gas production were measured using a gas chromatograph (GC; Infinicon 3000 microGC®) equipped with a thermal conductivity detector. A fixed molecular sieve 5A capillary column using argon as the carrier gas was used for hydrogen and nitrogen quantification, and a Plot U column with helium as the carrier gas was used for carbon dioxide quantification. GC data was processed using ezIQ and Diablo EZReporter software (Diablo Analytical, Antioch, CA, USA). For experiments performed in the anaerobic chamber, immediately after inoculation, 1.5 mL of each culture was transferred to a GC vial that was then sealed. For bioreactor sampling, the same parameters were utilized with samples automatically taken from the headspace every 11 min via an in-line setup. Gas production rates were derived from the GC gas percentage output using the ideal gas law, headspace flowrate (1 L min^{-1}), and the fermentation volume (1 L). The trapezoidal rule was used to approximate total gas production from the GC production rates using numerical integration.

3. Results

3.1. Co-culture Simulations Predict Higher Hydrogen Production than Both Monocultures

A consortium of *C. beijerinckii* NCIMB 8052 and *Y. regensburgei* ATCC 43,003 was originally identified as having the second highest predicted hydrogen production among 298,378 simulated co-cultures (Table 1) [39]. These co-culture simulations represented every pairwise combination of 773 human gut microbe GSMMs [38]. Upon further examination of output exchange fluxes for the co-culture of *C. beijerinckii* and *Y. regensburgei* and their monocultures, *C. beijerinckii* was predicted to be the sole hydrogen producer (Figure 2). Compared to the monoculture, the addition of *Y. regensburgei* boosted *C. beijerinckii* biomass (Figure 2) and hydrogen production (Figure 2). The increased hydrogen production was not due only to increased biomass, because when the hydrogen flux was normalized to growth rate, the co-culture hydrogen production rate (86 millimoles per gram dry cell weight per hour) was greater than the monoculture hydrogen production (69 millimoles per gram dry cell weight per hour). The increase in co-culture output fluxes were predicted to be due to cross-feeding of lactate from *Y. regensburgei* to *C. beijerinckii* (Figure 2). The overall ecological interaction was predicted to be commensal as the growth rate of *C. beijerinckii* increased from the monoculture to the co-culture, while the growth rate *Y. regensburgei* was unchanged (Figure 2). Additional microbial consortia containing *C. beijerinckii* that were predicted to have more than two-fold greater hydrogen production than *C. beijerinckii* alone are listed in Table 1.

Table 1. Microbial consortia predicted to have greater than two-fold *C. beijerinckii* hydrogen flux. Organisms predicted to increase the hydrogen flux greater than two-fold over the hydrogen flux of *C. beijerinckii* alone, when simulated in a co-culture with *C. beijerinckii*. Changes in predicted biomass for each organism (Δ Biomass Organism) and the predicted biomass of *C. beijerinckii* (Δ Biomass *C. beijerinckii*) from simulated co-cultures versus simulated monocultures.

Organism	Fold Hydrogen versus *C. Beijerinckii* Alone	Δ Biomass *C. Beijerinckii*	Δ Biomass Organism
Yokenella regensburgei ATCC 43003	2.25	0.58	0
Kluyvera ascorbata ATCC 33433	2.20	0.54	0
Hafnia alvei ATCC 51873	2.08	0.48	0
Yersinia rohdei ATCC 43380	2.07	0.49	0
Yersinia kristensenii ATCC 33638	2.07	0.49	0
Vibrio mimicus MB 451	2.06	0.43	0
Solobacterium moorei DSM 22971	2.06	−0.34	0.51
Escherichia hermannii NBRC 105704	2.05	0.44	0
Vibrio parahaemolyticus RIMD 2210633	2.05	0.45	0
Capnocytophaga sputigena ATCC 33612	2.05	0.43	0.21
Lactococcus garvieae ATCC 49156	2.05	−0.29	0.43
Trabulsiella guamensis ATCC 49490	2.01	0.42	0
Cellulosimicrobium cellulans J36	2.01	0.37	0.28

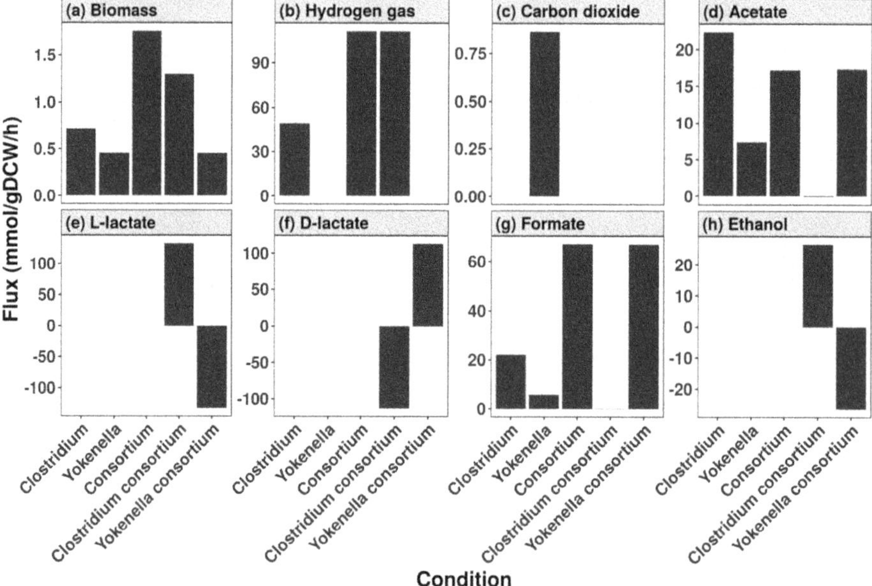

Figure 2. Predicted metabolite flux for a consortium of *C. beijerinckii* and *Y. regensburgei*. Flux balance analysis (FBA) simulations of *Clostridium beijerinckii* and *Yokenella regensburgei* monocultures (Alone), and co-cultures of *C. beijerinckii* and *Y. regensburgei* (Consortium) to predict production of (**a**) biomass, (**b**) hydrogen, (**c**) carbon dioxide, (**d**) acetate, (**e**) L-lactate, (**f**) D-lactate, (**g**) formate, (**h**) ethanol. *C. beijerinckii* Consortium indicates the contribution of *C. beijerinckii* to the metabolites in the co-culture FBA and *Y. regensburgei* Consortium indicates the contribution of *Y. regensburgei* to the metabolites in the co-culture FBA. Units of simulated flux are millimoles of metabolite produced per gram of dry cell weight per hour.

3.2. Metabolite Production by a Consortium of C. beijerinckii and Y. regensburgei

The consortium of *C. beijerinckii* NCIMB 8052 and *Y. regensburgei* ATCC 43,003 (Table 1) was predicted to exhibit the second largest increase in hydrogen production over any individual modeled bacterial strain when provided with a food waste simulant. To test for the potential of this interaction in vivo, cells were grown on an artificial garbage slurry medium (AGS) in an anaerobic chamber. After 24 h growth, *C. beijerinckii* and *Y. regensburgei* were plated on selective media to determine the number of colony forming units per milliliter of culture (CFU mL^{-1}) for each strain. We observed an increase in CFU mL^{-1} over the inoculum for both strains, indicating that AGS supports the growth of these bacteria independently (Figure 3a). Both strains also grew within 24 h in a co-culture, as measured by plating for CFUs on selective media (Figure 3a). The CFU count of each strain in the consortia was significantly lower than for each individual strain (Figure 3a). However, the total CFU count of both strains in the consortia was higher than for the cultures of *C. beijerinckii* alone and similar to the CFU count in the *Y. regensburgei* monocultures (Figure 3a). Viable cells were unable to be consistently recovered by plating for CFUs at subsequent times (data not shown), likely due to depletion of an essential growth factor or accumulation of a toxic byproduct in the media.

We measured gas production by *C. beijerinckii* and *Y. regensburgei*, either alone or as part of a consortium, as an increase in atmospheric percentage over the background atmosphere by gas chromatography (GC), after three days of growth on the artificial garbage slurry medium in sealed vials. *C. beijerinckii* produced approximately 8% hydrogen above the atmospheric background, whereas *Y. regensburgei* produced only about 1% hydrogen above the atmospheric background, when each was cultured individually (Figure 3b). The consortium produced a statistically similar percentage of hydrogen above the background ($p = 0.89$, by two-tailed Student's *t* test) to the *C. beijerinckii* monoculture (Figure 3b). Each individual strain and the consortium of the two strains produced an approximately equivalent proportional percentage of carbon dioxide to hydrogen (Figure 3b).

We quantified the organic acids produced by *C. beijerinckii* and *Y. regensburgei* individually and as a consortium from filtered extracts via HPLC. Both *C. beijerinckii* monocultures and the consortium primarily produced butyrate, consistent with previous reports [28], with the consortium producing a small but significantly larger amount than the *C. beijerinckii* monoculture (Figure 3c). As predicted by the FBA (Figure 2), the *Y. regensburgei* monoculture produced almost no butyrate (Figure 3c), but produced approximately 30 mM lactate, which was not observed in either the consortium or the *C. beijerinckii* monoculture (Figure 3d). The consortium produced substantially lower lactate than would be expected, if the *Y. regensburgei* present produced a proportional amount of lactate to that in the *Y. regensburgei* monoculture (Figure 3d). Acetate was a secondary fermentation product in all cultures, accumulating to approximately 11 mM in both the *C. beijerinckii* monoculture and the consortium, and approximately 4 mM in the *Y. regensburgei* monoculture (Figure 3e). A significant amount of formate was consumed in both monocultures, while the consortium produced a small amount of formate (Figure 3f). The consumption of formate by *Y. regensburgei* could be an indication of formate-hydrogen lyase activity [29], although low amounts of hydrogen were observed for *Y. regensburgei* monocultures (Figure 3b).

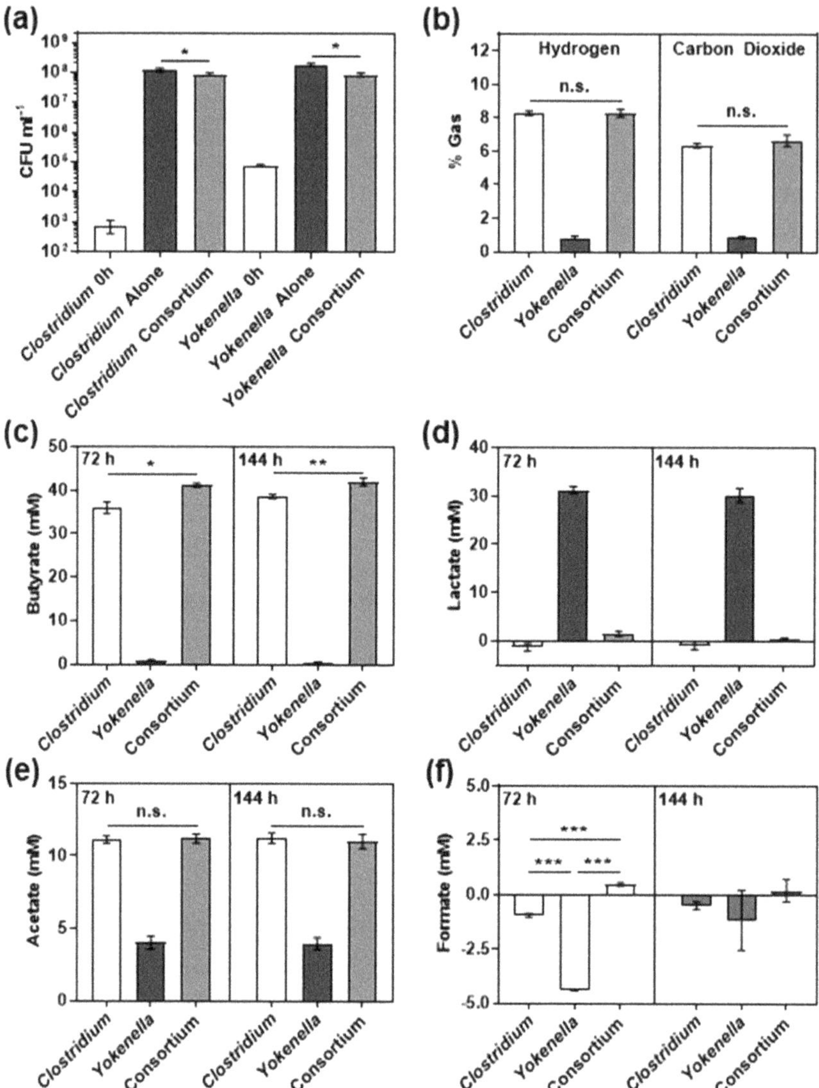

Figure 3. Metabolite production by a consortium of *C. beijerinckii* and *Y. regensburgei*. *Clostridium beijerinckii* and *Yokenella regensburgei* were grown as monocultures or as a consortium on artificial garbage slurry medium (AGS) for 144 h in an anaerobic chamber, either in gas chromatography vials (b) or sealed tubes (**a,c,d,e**). (**a**) Colony forming units after 24 h. (**b**) Normalized percent hydrogen and carbon dioxide measured by gas chromatography after 72 h. (**c**) Butyrate, (**d**) lactate, (**e**) acetate, and (**f**) formate production after 72 h and 144 h. Units for metabolites are mM as normalized to HPLC standards. The mean and standard error of the mean of four independent replicates are graphed. Asterisks indicate significant difference between the indicated samples (n.s. (not significant), $p > 0.05$; *, $p \leq 0.05$; **, $p \leq 0.01$; ***, $p \leq 0.001$ by two-tailed Student's *t* test).

3.3. Y. regensburgei Cross-feeds Lactate to C. beijerinckii for Butyrate Production

We hypothesized that the lactate produced by *Y. regensburgei* was utilized as a carbon source by *C. beijerinckii*, which subsequently produced additional butyrate, because a higher concentration of butyrate was observed in the consortium than in the *C. beijerinckii* monoculture (Figure 3c). To examine this possibility in a simpler system, we grew *C. beijerinckii* or *Y. regensburgei* on sterile filtered spent media from 24 h monocultures of *Y. regensburgei* or *C. beijerinckii* that were grown on Luria Bertani (LB) medium supplemented with 0.5% glucose. No growth was initially observed of either *C. beijerinckii* or *Y. regensburgei* on the sterile filtered spent media (data not shown); however, addition of sterile 50 mM potassium phosphate pH 6.8 to filtered spent media enabled growth of both strains (Figure 4a), suggesting that decreased media pH was inhibitory to growth. Growth of both *C. beijerinckii* and *Y. regensburgei* was significantly greater on spent media derived from the opposing species (*i.e.*, *C. beijerinckii* growth on *Y. regensburgei* spent media and *Y. regensburgei* growth on *C. beijerinckii* spent media) than observed for growth on spent media derived from the same species (i.e., *C. beijerinckii* growth on *C. beijerinckii* spent media and *Y. regensburgei* growth on *Y. regensburgei* spent media) (Figure 4a). This is consistent with the idea of cross-feeding between the two species.

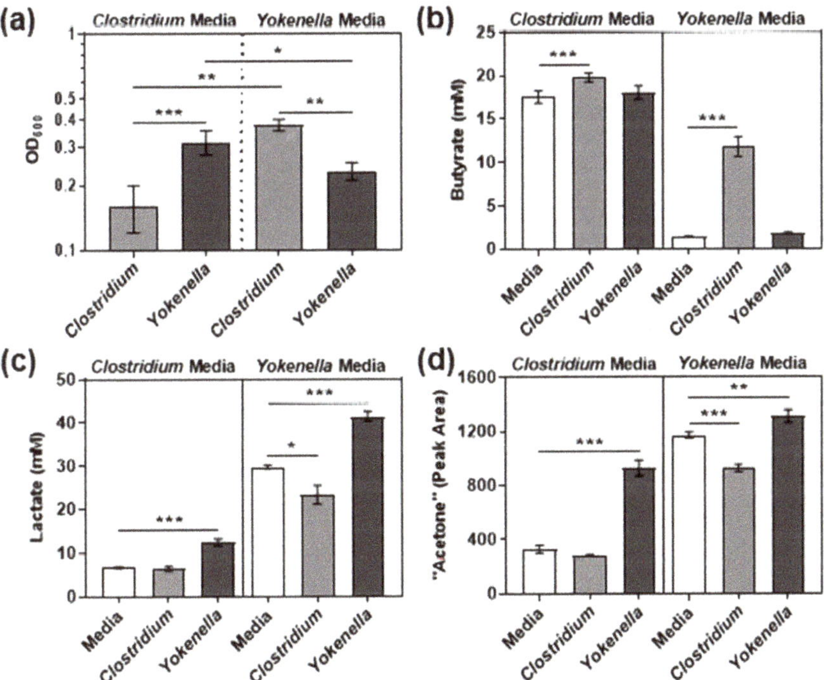

Figure 4. Spent media cross-feeding by *C. beijerinckii* and *Y. regensburgei*. *Clostridium beijerinckii* and *Yokenella regensburgei* were grown as monocultures for 24 h on LB + 0.5% glucose in an anaerobic chamber. Spent media was sterile filtered and 50 mM potassium phosphate pH 6.8 was added. Spent media from each species was inoculated 1:100 with overnight cultures of either *C. beijerinckii* or *Y. regensburgei* and after 24 h, (**a**) OD$_{600}$, (**b**) butyrate, (**c**) lactate, and (**d**) an unknown metabolite that elutes at the expected elution time for acetone, were measured. Units for metabolites are mM as normalized to HPLC standards. The mean and standard error of the mean of six independent replicates are graphed. Asterisks indicate significant difference between the indicated samples (*, $p \leq 0.05$; **, $p \leq 0.01$; ***, $p \leq 0.001$ by two-tailed Student's *t* test).

C. *beijerinckii* significantly increased the concentration of butyrate in spent media from both *C. beijerinckii* and *Y. regensburgei*, similar to growth on AGS; however, this increase was substantially larger in the *Y. regensburgei* spent media than the *C. beijerinckii* spent media (~700% versus ~12% increase) (Figure 4b). *C. beijerinckii* significantly depletes the lactate from *Y. regensburgei* spent media, concurrent with the large increase in butyrate (Figure 4c), suggesting that lactate produced by *Y. regensburgei* supports additional growth and butyrate production by *C. beijerinckii*. *Y. regensburgei* produces a significant amount of additional lactate in both *C. beijerinckii* and *Y. regensburgei* spent media (Figure 4c), suggesting that *C. beijerinckii* may produce nutrient(s) capable of being further metabolized by *Y. regensburgei*, and that pH is likely a limiting factor for its growth in LB supplemented with glucose.

The HPLC method used for metabolite detection was previously optimized for detection of short-chain fatty acids, acetone, butanol, and ethanol from *Clostridium acetobutylicum* [51]. A peak eluting at the time of the expected acetone peak was detected in the *Y. regensburgei* spent media (Figure 4d). *Y. regensburgei* has not previously been reported to produce acetone, and it does not encode the genes necessary to produce it [52]. The *C. beijerinckii* spent media also contains a metabolite with the same HPLC elution time; however, its peak area is only ~25–30% of the peak observed in the *Y. regensburgei* spent media. *C. beijerinckii* significantly depletes the undetermined metabolite when grown on the *Y. regensburgei* spent media (Figure 4d), while *Y. regensburgei* significantly increased the concentration of the metabolite when grown on both the *C. beijerinckii* and *Y. regensburgei* spent media (Figure 4d). Moreover, *Y. regensburgei* accumulates ~1.7-fold more of the metabolite when grown anaerobically than aerobically, despite aerobic cultures growing to more than two-fold higher optical density at 600 nm (data not shown). Electrospray ionization (ESI) mass spectrometry was performed on an HPLC fraction collected from the *Y. regensburgei* spent media corresponding to the acetone peak. The major mass to charge ratio (m/z) peaks that were not identified as solvent or carrier peaks were at m/z = 171.11 in the ESI+ spectrum and m/z = 127.99 and 111.00 in the ESI-spectrum (data not shown). These peaks do not map to a known metabolite that could plausibly be produced by *Y. regensburgei* given the current understanding of its metabolism. Further experimentation is required to determine the role of this molecule in potential cross-feeding between *Y. regensburgei* and *C. beijerinckii*.

3.4. C. beijerinckii Uses Lactate as a Carbon Source

Since *C. beijerinckii* is capable of depleting lactate from *Y. regensburgei* spent media (Figure 4c), and the GSMM predicted cross-feeding of lactate from *Y. regensburgei* to *C. beijerinckii* (Figure 2), we hypothesized that lactate could support the growth of *C. beijerinckii* either with or without a limited amount of glucose. The optical density of *C. beijerinckii* cultures indeed increased compared to an un-inoculated control in LB media containing 30 mM lactic acid and 20 mM potassium phosphate, when the initial pH was controlled to 6.7 with sodium hydroxide (Figure 5a). Concomitant with the increase in optical density, a significant amount of lactate was depleted from the media, suggesting its use as a carbon source by *C. beijerinckii* (Figure 5b). Moreover, *C. beijerinckii* grew (Figure 5a) and significantly depleted lactate from the media (Figure 5b) in cultures with 5 mM glucose and 15 mM lactate within 24 h, although the amount of growth was significantly less than observed in cultures containing 10 mM glucose during the same duration (Figure 5a). In both cases, glucose was completely depleted from the media (data not shown). *C. beijerinckii* produced primarily butyrate in each of the three conditions, with the amount produced positively correlating with the amount of glucose present (Figure 5c). *C. beijerinckii* produced substantially more ethanol in the absence of lactate, exhibiting low ethanol production in both the lactate alone and mixed carbon source cultures (Figure 5d).

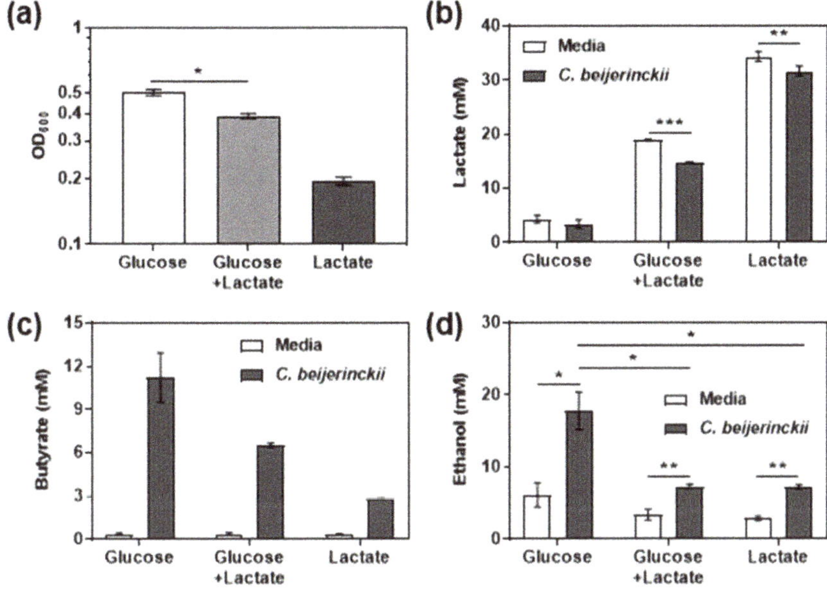

Figure 5. Lactate can support the growth of *C. beijerinckii* with or without glucose. *C. beijerinckii* was grown on LB + 20 mM potassium phosphate pH 6.8 + either 10 mM glucose, or 30 mM lactate, or 5 mM glucose and 15 mM lactate, in an anaerobic chamber. (**a**) OD_{600}, (**b**) lactate, (**c**) butyrate, and (**d**) ethanol, were measured after 24 h from the *C. beijerinckii* cultures and un-inoculated controls (Media). Units for metabolites are mM as normalized to HPLC standards. The mean and standard error of the mean of four independent replicates are graphed. Asterisks indicate significant difference between the indicated samples (n.s., *, $p > 0.05$; **, $p \leq 0.01$; ***, $p \leq 0.001$ by two-tailed Student's *t* test).

3.5. Enhanced Metabolite Production by the Consortium Is Condition-Dependent

To quantify the production rate and amount of gases produced by *C. beijerinckii* and *Y. regensburgei* either alone or as a consortium, we used a DASGIP® parallel bioreactor system coupled with two gas chromatographs to perform near constant measurements of gases (every 11 min) from two independent replicates grown in one liter volumes on AGS. The experiment was performed with two distinct sets of bioreactor conditions. First, to closely mimic the closed tubes in the anaerobic chamber, the bioreactor mixtures were left stationary until fifteen seconds prior to sample collection and pH was not controlled. Second, to maintain the optimal pH for hydrogen production by the *C. beijerinckii* hydrogenase [25], the pH of bioreactor vessels was automatically controlled at pH 6.7, with continuous stirring at 100 rpm for the entirety of the experiment.

The two bioreactor experiments resulted in distinct patterns of growth and metabolite production for both the consortium and monocultures of *C. beijerinckii* and *Y. regensburgei*. In the stationary bioreactor experiments without pH control, *C. beijerinckii* grew to a higher density in monoculture than the consortium starting eight hours after inoculation and continuing for the duration of the experiment, until it was undetectable in both the monoculture and consortia by the 48 h time point (Figure 6a). By contrast, no difference was observed in the amount of colony forming units for *Y. regensburgei* between the monoculture and the consortium until the 36 h time point when there was a 2-log decrease of *Y. regensburgei* in the consortium. *Y. regensburgei* levels further decreased to undetectable levels by the 48 h time point (Figure 6a). Although a lower initial amount of *C. beijerinckii* was present in the consortium, both *C. beijerinckii* and *Y. regensburgei* were present in similar amounts from 12–24 h (Figure 6a). With pH control, *Y. regensburgei* maintained a high level of culturable CFUs in both the monoculture and the consortium, even at 72 h (Figure 7a). With pH control, *C. beijerinckii* in both monoculture and the consortium reached an apparent peak density at 12 h, before decreasing to nearly undetectable levels by 72 h (Figure 7a).

C. beijerinckii monocultures and the consortium produced similar amounts of hydrogen (Figure 6b) and carbon dioxide (Figure 6c), without pH control, though variability between samples was observed. With pH control, *C. beijerinckii* monocultures produced greater amounts of hydrogen (Figure 7b) and carbon dioxide (Figure 7c) than the consortium. The total amounts of hydrogen and carbon dioxide produced were higher for fermentations with pH control than without, although differences in the kinetics between the individual bioreactors of both *C. beijerinckii* monocultures and the consortium with pH control were observed. *Y. regensburgei* produced small amounts of hydrogen (Figures 6b and 7b) and carbon dioxide (Figures 6c and 7c) throughout the experiments with and without pH control.

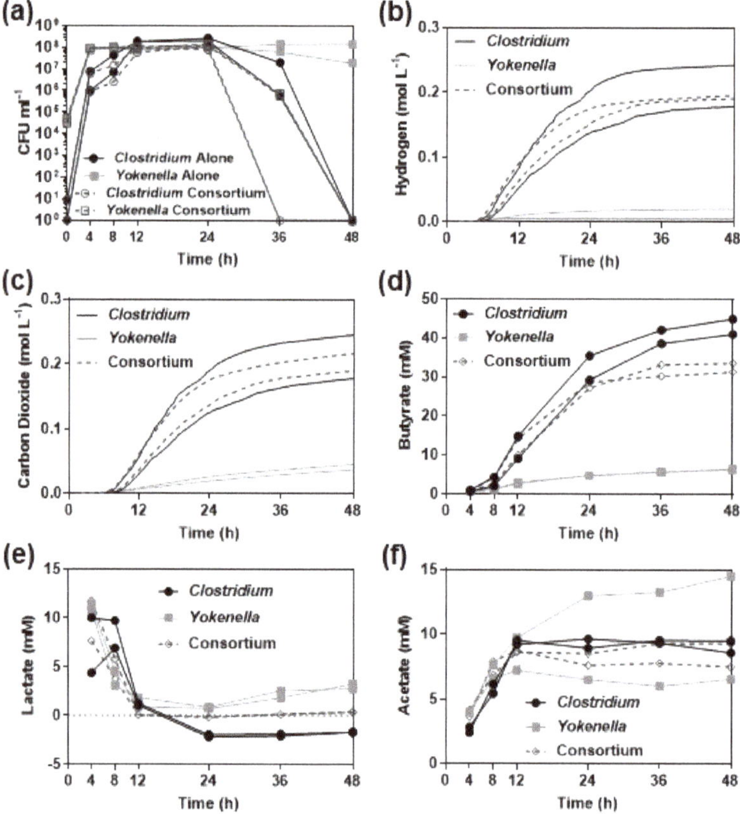

Figure 6. Metabolites produced by the *C. beijerinckii* and *Y. regensburgei* consortium in stationary bioreactors. *Clostridium beijerinckii* and *Yokenella regensburgei* were grown as monocultures or as a consortium on AGS for 48 h in 1 L bioreactor vessels. Vessel contents were only stirred at 100 rpm for 15 s prior to collection of each sample. (**a**) Colony forming units, (**b**) hydrogen, (**c**) carbon dioxide, (**d**) butyrate, (**e**) lactate, (**f**) acetate quantified at the indicated time points. Each replicate is graphed independently. Units for gas production are moles of gas produced per liter of fermentation volume (mol L^{-1}), derived from the GC gas percentage output using the ideal gas law, headspace flowrate, and the fermentation volume. Units for metabolites are mM as normalized to HPLC standards.

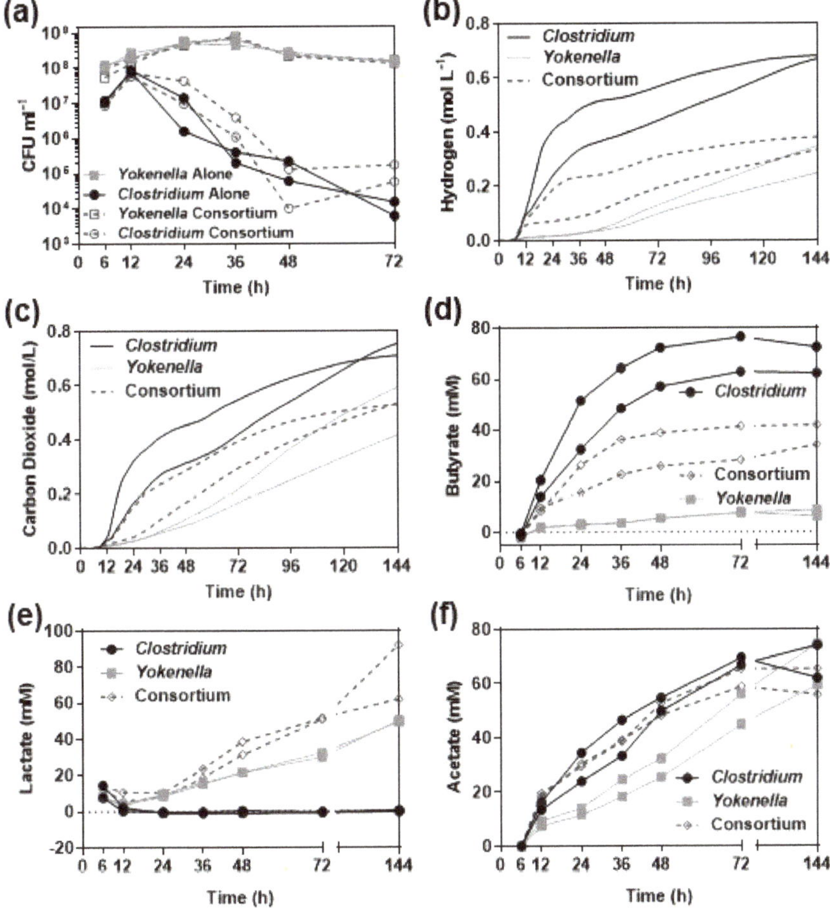

Figure 7. Metabolites produced by the *C. beijerinckii* and *Y. regensburgei* consortium in bioreactors with stirring and pH control. *Clostridium beijerinckii* and *Yokenella regensburgei* were grown as monocultures or as a consortium on AGS for 144 h in 1 L bioreactor vessels, with the pH maintained at 6.7 via the addition of sodium hydroxide and constant stirring at 100 rpm. (**a**) Colony forming units, (**b**) hydrogen, (**c**) carbon dioxide, (**d**) butyrate, (**e**) lactate, (**f**) acetate quantified at the indicated time points. Each replicate is graphed independently. Units for gas production are moles of gas produced per liter of fermentation volume (mol L^{-1}), derived from the GC gas percentage output using the ideal gas law, headspace flowrate, and the fermentation volume. Units for metabolites are mM as normalized to HPLC standards.

C. beijerinckii monocultures produced primarily butyrate to a higher level than the consortium both with pH control (Figure 7d) and without pH control (Figure 6d). *Y. regensburgei* monocultures produced primarily lactate in both conditions (Figures 6e and 7e). However, the amount of lactate produced by *Y. regensburgei* with pH control was lower than the amount of lactate produced by the consortium (Figure 7e), and the amount of lactate produced by *Y. regensburgei* without pH control was substantially lower than in any other experiment (Figure 6e). *C. beijerinckii*, *Y. regensburgei*, and the consortium produced similar levels of acetate (Figures 6f and 7f) and ethanol (data not shown) as secondary fermentation products in both the experiments with and without pH control.

4. Discussion

Anaerobic fermentation of food waste is an attractive alternative to other disposal methods, to reduce waste volume and recover lost energy [1–3]. Here, we developed a framework to experimentally test metabolic modeling predictions of microbial consortia for their ability to produce commodity chemicals (Figure 1). We have established that a consortium of *C. beijerinckii* and *Y. regensburgei* is capable of producing increased amounts of commodity chemicals from a simulated food waste medium, compared to monocultures of either species. Genome-scale metabolic modeling predicted that co-cultures of *C. beijerinckii* and *Y. regensburgei* would result in synergistic overproduction of hydrogen as a result of lactate cross-feeding from *Y. regensburgei* to *C. beijerinckii*. We observed similar hydrogen gas production by *C. beijerinckii* and the consortium in stationary conditions (Figures 3b and 6b) and overproduction of the commodity chemicals butyrate (Figure 3c) and lactate (Figure 7e) by the consortium in distinct experimental conditions. Moreover, we demonstrated that *C. beijerinckii* is capable of using exogenous lactate as a carbon source (Figure 5) and that *Y. regensburgei* cross-feeds lactate to *C. beijerinckii* (Figure 4). Differences in the metabolites produced by the consortium of *C. beijerinckii* and *Y. regensburgei* during experiments with the artificial garbage slurry medium (Figures 3, 6 and 7) exhibit the difficulty in fermentation scale-up, but also demonstrate that opportunities exist to modulate fermentation conditions for varied commodity chemical output.

4.1. Experimental Implementation of Genome-Scale Metabolic Modeling Predictions

The consortium of *C. beijerinckii* and *Y. regensburgei* was predicted by GSMM and FBA to have the second highest overproduction of hydrogen gas compared to any individual species from models of 773 gut microbiota species [38,39], leading to its selection for further experimentation. Similar levels of hydrogen were observed for *C. beijerinckii* and the consortium of *C. beijerinckii* and *Y. regensburgei* in small-scale experiments performed in an anaerobic chamber (Figure 3). This experimental setup prevented quantification of gas from more than a single time point. GC vials from the experiment all had the same headspace volume, but slight variations in growth may have caused different pressures for any given vial at time of headspace analysis. It is possible to measure the pressure in a given vial; however, this was not done, as it is difficult to obtain a pressure measurement without releasing gas from the GC vial, which could potentially affect the GC concentration measurements. Continuous GC monitoring over time was performed from 1 L bioreactor cultures (Figures 6 and 7). For stationary cultures without pH control, which most closely mimicked the small-scale experimental design, the hydrogen production by the consortium was similar to the amount of hydrogen produced by *C. beijerinckii* monocultures (Figure 6b). While the consortium did not produce similar amounts of hydrogen to *C. beijerinckii* monocultures grown with pH control, which exhibited the highest amounts of hydrogen produced from any experimental condition tested (Figure 7b), the FBA modeling did not account for such conditions. Manipulating simulated conditions with FBA modeling could provide additional avenues to improve commodity chemical production.

4.2. Y. regensburgei Cross-feeding to C. beijerinckii

In small-scale fermentations of the AGS, substantially lower lactate was produced by the consortium than would be expected if the *Y. regensburgei* present produced a proportional amount of lactate to that in the *Y. regensburgei* monoculture (Figure 3d). This suggested that the predicted cross-feeding of lactate from *Y. regensburgei* to *C. beijerinckii* (Figure 2) may be occurring when the consortium was grown in AGS. *Clostridium butyricum* [53] and *Clostridium saccharoperbutylacetonicum* [54] were previously described to metabolize acetate and lactate to butyrate, while several other *Clostridium*, including *C. beijerinckii*, were proposed to breakdown lactate [55]. Here, we show that *C. beijerinckii* can metabolize a limited amount of lactate, with or without a limited amount of glucose (Figure 5). The proposed mechanism for lactate metabolism in *Clostridium* is based on the lactate oxidation pathway in *Acetobacterium woodii*, which couples a flavin adenine dinucleotide (FAD)-dependent lactate dehydrogenase with an electron flavoprotein complex to convert a reduced ferredoxin, lactate, and two oxidized nicotinamide adenine dinucleotides (NAD) to an oxidized ferredoxin, pyruvate, and two reduced nicotinamide adenine dinucleotides (NADH) [53]. A similar mechanism is likely used by *C. beijerinckii*, as a genomic locus with significant sequence homology to the locus proposed to breakdown lactate in *C. butyricum* [53] is found in *C. beijerinckii* (*Cbei_2884-Cbei_2888*) using the Basic Local Alignment Search Tool (BLAST) [56]. While FBA predicts that multiple species can cross-feed lactate to *C. beijerinckii*, the ability to cross-feeding of lactate to *Clostridium* may be somewhat limited, as lactic acid bacteria, which produce large amounts of lactate, are often detrimental to efforts to produce hydrogen from complex microbial consortia [55]. Moreover, a pilot experiment attempting to cross-feed lactate from the lactic acid bacterium *Lactobacillus fermentum* to *C. beijerinckii* was unsuccessful, as *L. fermentum* produced a significant amount of lactate (~30 mM), but no change in lactate or butyrate levels was observed when *C. beijerinckii* was cultured in the spent medium (data not shown).

4.3. Varying Growth Conditions to Control Metabolic Output

While small variability in metabolic output was observed between replicates with small-scale cultures (Figure 3), a substantial amount of variability was observed between all individual replicates containing *C. beijerinckii* for the bioreactor experiments (Figures 6 and 7). Part of this variability may stem from differences in handling of the inoculums for the bioreactors versus the small-scale cultures. *C. beijerinckii* is sensitive to oxygen, and inoculation of the bioreactors required brief exposure to oxygen as the inoculum was transported from an anaerobic chamber to the anaerobic environment of the bioreactors. This exposure coupled with slight differences in the growth phase or culture density of the *C. beijerinckii* inocula could have biased individual bioreactors to be more or less favorable for *C. beijerinckii* growth. The AGS is another potential source of variability, as the precise composition of the main component could have varied from vessel to vessel. The pH control also could have introduced variability between replicates, because noticeably different amounts of sodium hydroxide were automatically added to each vessel, which differentially affected the pH and osmolality of the medium. Both pH and sodium concentration can have profound effects on *Clostridium* metabolic output [57,58].

In the experiments with pH control, the consortium of *C. beijerinckii* and *Y. regensburgei* produced a drastically different metabolic profile than in other experiments (Figure 7). After an initial lactate decrease, which corresponded to an increase in *C. beijerinckii* colony forming units (Figures 7a and 6e), the consortium cultures produced more lactate than *Y. regensburgei* monocultures. This increase in lactate corresponded to a decrease in *C. beijerinckii*, while *Y. regensburgei* levels remained near constant. Thus, *Y. regensburgei* likely dominated the metabolite profile of the cultures, and *C. beijerinckii* may have either shifted its metabolism to produce additional lactate or cross-fed a metabolite to *Y. regensburgei* increasing its lactate production. Modifying the metabolic output of a consortium by either controlling or not controlling the culture pH is a mechanism that could be used to increase the agility of a designed microbial consortium for a future application.

4.4. Microbial Consortia with Distinct Mechanisms of Chemical Overproduction

Modeling predicted the consortium of *C. beijerinckii* and *Y. regensburgei* to produce increased *C. beijerinckii* biomass (Table 1), due to the cross-feeding of lactate from *Y. regensburgei* to *C. beijerinckii* that was experimentally observed (Figure 4). Other consortia that were predicted to produce an increased amount of hydrogen compared to *C. beijerinckii* monocultures (Table 1) likely have different mechanisms that could contribute to commodity chemical overproduction. For example, *Cellulosimicrobium cellulans* can break down cellulose and xylans [59], complex polysaccharides common in plant material [60], which are inaccessible to *C. beijerinckii*. In contrast to the consortium of *C. beijerinckii* and *Y. regensburgei*, FBA predicted that the biomass of both *C. beijerinckii* and *Cellulosimicrobium cellulans* would increase in a co-culture compared to monocultures (Table 1), perhaps due to increased available carbon and cross-feeding of metabolites from *C. beijerinckii* to *Cellulsimicrobium cellulans*. Similarly, members of the *Capnocytophaga* genus can metabolize complex polysaccharides [61] and carbon dioxide [62], which could explain the predicted hydrogen overproduction by a consortium of *C. beijerinckii* and *Capnocytophaga sputigena* (Table 1). Integrating genome-scale metabolic model improvements and testing additional consortia with different predicted interactions could enable discovery of pathways to enhanced commodity chemical production.

5. Conclusions

Genome-scale metabolic models and flux-balance analysis predicted several microbial consortia expected to produce a significantly greater amount of hydrogen than any individual species [39] (Table 1). We focused on the consortium of *C. beijerinckii* and *Y. regensburgei* that was predicted to have the greatest increase in biomass and hydrogen production by *C. beijerinckii*. The flux-balance analysis predicted the increase in hydrogen and biomass based on the cross-feeding of lactate to *C. beijerinckii* (Figure 2). We established that cross-feeding of lactate from *Y. regensburgei* to *C. beijerinckii* can occur (Figure 4), and that exogenous lactate is capable of supporting the growth of *C. beijerinckii* (Figure 5). We were unable to demonstrate that growing *C. beijerinckii* as part of a consortium with *Y. regensburgei* increased the production of hydrogen above the levels observed for *C. beijerinckii* monocultures (Figure 3). The consortium was capable of producing more butyrate (Figure 3c) or lactate (Figure 7e) than individual monocultures, depending on the growth conditions. Examining the effects of adding another species that is predicted to have a different role in promoting hydrogen production than cross-feeding of lactate to the consortium of *C. beijerinckii* and *Y. regensburgei*, or testing that species with *C. beijerinckii* alone could provide better improvement of hydrogen production from the artificial garbage slurry than was observed. The framework presented here can be used to screen large numbers of possible microbial combinations by first using genome-scale metabolic modeling and flux-balance analysis to predict consortia with an increased likelihood to convert waste to commodity chemicals and then testing the highest producing consortia experimentally for their ability to produce chemicals of interest from readily available materials.

Author Contributions: N.D.S., M.A.P., and K.L.A. conceptualized the research; M.A.P. performed the GSMM and FBA, and wrote the original draft of the Introduction, Materials and Methods, and Results section for the GSMM. W.M. performed the HPLC and GC analysis and wrote the original draft of the Materials and Methods section for the HPLC and GC. N.D.S. and E.S.G. performed the bioreactor fermentation experiments and edited the paper. M.B. performed the analysis of the bioreactor GC data. N.D.S. performed all other experiments and wrote the original draft of the remainder of the paper.

Funding: This research received no external funding.

Acknowledgments: The authors would like to acknowledge members of the Combat Capabilities and Development Command Army Research Laboratory for useful discussions and Yue Li of the University of Maryland for performing mass spectrometry.

Conflicts of Interest: The authors declare no conflict of interest. The funders had no role in the design of the study; in the collection, analyses, or interpretation of data; in the writing of the manuscript, or in the decision to publish the results.

References

1. Xu, F.; Li, Y.; Ge, X.; Yang, L.; Li, Y. Anaerobic digestion of food waste—Challenges and opportunities. *Bioresour. Technol.* **2018**, *247*, 1047–1058. [CrossRef] [PubMed]
2. Ren, Y.; Yu, M.; Wu, C.; Wang, Q.; Gao, M.; Huang, Q.; Liu, Y. A comprehensive review on food waste anaerobic digestion: Research updates and tendencies. *Bioresour. Technol.* **2018**, *247*, 1069–1076. [CrossRef] [PubMed]
3. Ma, Y.; Liu, Y. Turning food waste to energy and resources towards a great environmental and economic sustainability: An innovative integrated biological approach. *Biotechnol. Adv.* **2019**. [CrossRef] [PubMed]
4. Rock, K.; Lesher, L.; Kramer, F.M.; Johnson, J.; Bordic, M.; Miller, H. An Analysis of Military Field Feeding Waste. In *U. S. Army Natick Soldier Systems Center Technology Report Natick/Tr-00/021*; U.S. Army Natick Soldier Systems Center: Natick, MA, USA, 2000.

5.   Knowlton, L.; Pickard, D.; Diebold, J.; Lasnik, L.; Lilley, A.; Browne, K. On-Site Field-Feeding Waste to Energy Converter. 2008. Available online: http://nsrdeec.natick.army.mil/LIBRARY/00-09/R08-105.pdf (accessed on 15 August 2019).

6.   Birgen, C.; Durre, P.; Preisig, H.A.; Wentzel, A. Butanol production from lignocellulosic biomass: Revisiting fermentation performance indicators with exploratory data analysis. *Biotechnol. Biofuels* **2019**, *12*, 167. [CrossRef] [PubMed]

7.   Cairns, T.C.; Zheng, X.; Zheng, P.; Sun, J.; Meyer, V. Moulding the mould: Understanding and reprogramming filamentous fungal growth and morphogenesis for next generation cell factories. *Biotechnol. Biofuels* **2019**, *12*, 77. [CrossRef]

8.   Shin, H.D.; McClendon, S.; Vo, T.; Chen, R.R. *Escherichia coli* binary culture engineered for direct fermentation of hemicellulose to a biofuel. *Appl. Environ. Microbiol.* **2010**, *76*, 8150–8159. [CrossRef]

9.   Zhang, H.; Pereira, B.; Li, Z.; Stephanopoulos, G. Engineering *Escherichia coli* coculture systems for the production of biochemical products. *Proc. Natl. Acad. Sci. USA* **2015**, *112*, 8266–8271. [CrossRef]

10.   Salmela, M.; Lehtinen, T.; Efimova, E.; Santala, S.; Mangayil, R. Metabolic pairing of aerobic and anaerobic production in a one-pot batch cultivation. *Biotechnol. Biofuels* **2018**, *11*, 187. [CrossRef]

11.   Kim, H.; Jeon, B.S.; Pandey, A.; Sang, B.I. New coculture system of *Clostridium spp.* and *Megasphaera hexanoica* using submerged hollow-fiber membrane bioreactors for caproic acid production. *Bioresour. Technol.* **2018**, *270*, 498–503. [CrossRef]

12.   Wen, Z.; Minton, N.P.; Zhang, Y.; Li, Q.; Liu, J.; Jiang, Y.; Yang, S. Enhanced solvent production by metabolic engineering of a twin-clostridial consortium. *Metab. Eng.* **2017**, *39*, 38–48. [CrossRef]

13.   Gomez-Flores, M.; Nakhla, G.; Hafez, H. Hydrogen production and microbial kinetics of *Clostridium termitidis* in mono-culture and co-culture with *Clostridium beijerinckii* on cellulose. *AMB Express* **2017**, *7*, 84. [CrossRef] [PubMed]

14.   Roell, G.W.; Zha, J.; Carr, R.R.; Koffas, M.A.; Fong, S.S.; Tang, Y.J. Engineering microbial consortia by division of labor. *Microb. Cell Fact.* **2019**, *18*, 35. [CrossRef]

15.   Zhou, K.; Qiao, K.; Edgar, S.; Stephanopoulos, G. Distributing a metabolic pathway among a microbial consortium enhances production of natural products. *Nat. Biotechnol.* **2015**, *33*, 377–383. [CrossRef] [PubMed]

16.   Cavaliere, M.; Feng, S.; Soyer, O.S.; Jimenez, J.I. Cooperation in microbial communities and their biotechnological applications. *Environ. Microbiol.* **2017**, *19*, 2949–2963. [CrossRef] [PubMed]

17.   LaSarre, B.; McCully, A.L.; Lennon, J.T.; McKinlay, J.B. Microbial mutualism dynamics governed by dose-dependent toxicity of cross-fed nutrients. *ISME J.* **2017**, *11*, 337–348. [CrossRef]

18.   Charubin, K.; Papoutsakis, E.T. Direct cell-to-cell exchange of matter in a synthetic *Clostridium* syntrophy enables CO2 fixation, superior metabolite yields, and an expanded metabolic space. *Metab. Eng.* **2019**, *52*, 9–19. [CrossRef]

19.   Benomar, S.; Ranava, D.; Cardenas, M.L.; Trably, E.; Rafrafi, Y.; Ducret, A.; Hamelin, J.; Lojou, E.; Steyer, J.P.; Giudici-Orticoni, M.T. Nutritional stress induces exchange of cell material and energetic coupling between bacterial species. *Nat. Commun.* **2015**, *6*, 7283. [CrossRef]

20.   Kosina, S.M.; Danielewicz, M.A.; Mohammed, M.; Ray, J.; Suh, Y.; Yilmaz, S.; Singh, A.K.; Arkin, A.P.; Deutschbauer, A.M.; Northen, T.R. Exometabolomics Assisted Design and Validation of Synthetic Obligate Mutualism. *ACS Synth. Biol.* **2016**, *5*, 569–576. [CrossRef]

21.   Smith, N.W.; Shorten, P.R.; Altermann, E.; Roy, N.C.; McNabb, W.C. The Classification and Evolution of Bacterial Cross-Feeding. *Front. Ecol. Evol.* **2019**, *7*. [CrossRef]

22.   Łukajtis, R.; Hołowacz, I.; Kucharska, K.; Glinka, M.; Rybarczyk, P.; Przyjazny, A.; Kamiński, M. Hydrogen production from biomass using dark fermentation. *Renew. Sustain. Energy Rev.* **2018**, *91*, 665–694. [CrossRef]

23.   Keskin, T.; Abo-Hashesh, M.; Hallenbeck, P.C. Photofermentative hydrogen production from wastes. *Bioresour. Technol.* **2011**, *102*, 8557–8568. [CrossRef] [PubMed]

24.   Yoo, M.; Bestel-Corre, G.; Croux, C.; Riviere, A.; Meynial-Salles, I.; Soucaille, P. A Quantitative System-Scale Characterization of the Metabolism of *Clostridium Acetobutylicum*. *mBio* **2015**, *6*, 1808–1815. [CrossRef] [PubMed]

25.   Masset, J.; Calusinska, M.; Hamilton, C.; Hiligsmann, S.; Joris, B.; Wilmotte, A.; Thonart, P. Fermentative hydrogen production from glucose and starch using pure strains and artificial co-cultures of *Clostridium Spp.* *Biotechnol. Biofuels* **2012**, *5*, 35. [CrossRef] [PubMed]

26. Liu, B.F.; Ren, N.Q.; Xie, G.J.; Ding, J.; Guo, W.Q.; Xing, D.F. Enhanced bio-hydrogen production by the combination of dark- and photo-fermentation in batch culture. *Bioresour. Technol.* **2010**, *101*, 5325–5329. [CrossRef] [PubMed]

27. Laurinavichene, T.; Laurinavichius, K.; Shastik, E.; Tsygankov, A. Long-term H_2 photoproduction from starch by co-culture of *Clostridium butyricum* and *Rhodobacter sphaeroides* in a repeated batch process. *Biotechnol. Lett.* **2018**, *40*, 309–314. [CrossRef]

28. Nasr, N.; Gupta, M.; Hafez, H.; El Naggar, M.H.; Nakhla, G. Mono- and co-substrate utilization kinetics using mono- and co-culture of *Clostridium beijerinckii* and *Clostridium saccharoperbutylacetonicum*. *Bioresour. Technol.* **2017**, *241*, 152–160. [CrossRef]

29. Maeda, T.; Tran, K.T.; Yamasaki, R.; Wood, T.K. Current state and perspectives in hydrogen production by *Escherichia coli*: Roles of hydrogenases in glucose or glycerol metabolism. *Appl. Microbiol. Biotechnol.* **2018**, *102*, 2041–2050. [CrossRef]

30. Seppälä, J.J.; Puhakka, J.A.; Yli-Harja, O.; Karp, M.T.; Santala, V. Fermentative hydrogen production by *Clostridium butyricum* and *Escherichia coli* in pure and cocultures. *Int. J. Hydrogen Energy* **2011**, *36*, 10701–10708. [CrossRef]

31. Schellenberger, J.; Que, R.; Fleming, R.M.; Thiele, I.; Orth, J.D.; Feist, A.M.; Zielinski, D.C.; Bordbar, A.; Lewis, N.E.; Rahmanian, S.; et al. Quantitative prediction of cellular metabolism with constraint-based models: The COBRA Toolbox v2.0. *Nat. Protoc.* **2011**, *6*, 1290–1307. [CrossRef]

32. Orth, J.D.; Thiele, I.; Palsson, B.O. What is flux balance analysis? *Nat. Biotechnol.* **2010**, *28*, 245–248. [CrossRef]

33. McCloskey, D.; Palsson, B.O.; Feist, A.M. Basic and applied uses of genome-scale metabolic network reconstructions of *Escherichia coli*. *Mol. Syst. Biol.* **2013**, *9*, 661. [CrossRef] [PubMed]

34. Zhuang, K.; Izallalen, M.; Mouser, P.; Richter, H.; Risso, C.; Mahadevan, R.; Lovley, D.R. Genome-scale dynamic modeling of the competition between *Rhodoferax* and *Geobacter* in anoxic subsurface environments. *ISME J.* **2011**, *5*, 305–316. [CrossRef] [PubMed]

35. Stolyar, S.; Van Dien, S.; Hillesland, K.L.; Pinel, N.; Lie, T.J.; Leigh, J.A.; Stahl, D.A. Metabolic modeling of a mutualistic microbial community. *Mol. Syst. Biol.* **2007**, *3*, 92. [CrossRef] [PubMed]

36. Bauer, E.; Zimmermann, J.; Baldini, F.; Thiele, I.; Kaleta, C. BacArena: Individual-based metabolic modeling of heterogeneous microbes in complex communities. *PLoS Comput. Biol.* **2017**, *13*, e1005544. [CrossRef] [PubMed]

37. Harcombe, W.R.; Riehl, W.J.; Dukovski, I.; Granger, B.R.; Betts, A.; Lang, A.H.; Bonilla, G.; Kar, A.; Leiby, N.; Mehta, P.; et al. Metabolic resource allocation in individual microbes determines ecosystem interactions and spatial dynamics. *Cell Rep.* **2014**, *7*, 1104–1115. [CrossRef] [PubMed]

38. Magnusdottir, S.; Heinken, A.; Kutt, L.; Ravcheev, D.A.; Bauer, E.; Noronha, A.; Greenhalgh, K.; Jager, C.; Baginska, J.; Wilmes, P.; et al. Generation of genome-scale metabolic reconstructions for 773 members of the human gut microbiota. *Nat. Biotechnol.* **2016**. [CrossRef]

39. Perisin, M.A.; Sund, C.J. Human gut microbe co-cultures have greater potential than monocultures for food waste remediation to commodity chemicals. *Sci. Rep.* **2018**, *8*, 15594. [CrossRef]

40. R Core Team. R: A language and environment for statistical computing. In *R Foundation for Statistical Computing*; R Core Team: Vienna, Austria, 2015; Available online: http://www.R-project.org/ (accessed on 24 September 2019).

41. Wickham, H. *Ggplot2: Elegant Graphics for Data Analysis*; Springer: New York, NY, USA, 2009.

42. Fritzemeier, C.J.; Gelius-Dietrich, G.; Alzoubi, D.; Habil, A. sybilSBML: SBML Integration in Package 'Sybil'. 2017. Available online: https://cran.r-project.org/web/packages/sybilSBML/index.html (accessed on 24 September 2019).

43. Gelius-Dietrich, G.; Desouki, A.A.; Fritzemeier, C.J.; Lercher, M.J. Sybil—Efficient constraint-based modelling in R. *BMC Syst. Biol.* **2013**, *7*, 125. [CrossRef]

44. Fritzemeier, C.J.; Gelius-Dietrich, G.; Luangkesorn, L. glpkAPI: R Interface to C API of GLPK. 2015. Available online: https://cran.r-project.org/web/packages/glpkAPI/index.html (accessed on 24 September 2019).

45. Heirendt, L.; Arreckx, S.; Pfau, T.; Mendoza, S.N.; Richelle, A.; Heinken, A.; Haraldsdottir, H.S.; Keating, S.M.; Vlasov, V.; Wachowiak, J.; et al. Creation and analysis of biochemical constraint-based models: The COBRA Toolbox v3.0. *Nat. Protoc* **2019**, *14*, 639–702. [CrossRef]

46. Klitgord, N.; Segre, D. Environments that Induce Synthetic Microbial Ecosystems. *PLoS Comput. Biol.* **2010**, *6*, 1001002. [CrossRef]

47.	Roos, J.W.; McLaughlin, J.K.; Papoutsakis, E.T. The effect of pH on nitrogen supply, cell lysis, and solvent production in fermentations of *Clostridium acetobutylicum*. *Biotechnol. Bioeng.* **1985**, *27*, 681–694. [CrossRef] [PubMed]

48.	O'Brien, R.W.; Morris, J.G. The Ferredoxin-dependent reduction of chloramphenicol by clostridium acetobutylicum. *J. Gen. Microbiol.* **1971**, *67*, 265–271. [CrossRef] [PubMed]

49.	Stock, I.; Sherwood, K.J.; Wiedemann, B. Antimicrobial susceptibility patterns, beta-lactamases, and biochemical identification of *Yokenella regensburgei* strains. *Diagn. Microbiol. Infect. Dis.* **2004**, *48*, 5–15. [CrossRef] [PubMed]

50.	Sasaki, K.; Morita, M.; Hirano, S.; Ohmura, N.; Igarashi, Y. Effect of adding carbon fiber textiles to methanogenic bioreactors used to treat an artificial garbage slurry. *J. Biosci. Bioeng.* **2009**, *108*, 130–135. [CrossRef] [PubMed]

51.	Finch, A.S.; Mackie, T.D.; Sund, C.J.; Sumner, J.J. Metabolite analysis of *Clostridium acetobutylicum*: Fermentation in a microbial fuel cell. *Bioresour. Technol.* **2011**, *102*, 312–315. [CrossRef] [PubMed]

52.	Kosako, Y.; Sakazaki, R.; Yoshizaki, E. Yokenella regensburgei gen. nov., sp. nov.: A new genus and species in the family Enterobacteriaceae. *Jpn. J. Med. Sci. Biol.* **1984**, *37*, 117–124. [CrossRef]

53.	Detman, A.; Mielecki, D.; Chojnacka, A.; Salamon, A.; Błaszczyk, M.K.; Sikora, A. Cell factories converting lactate and acetate to butyrate: *Clostridium butyricum* and microbial communities from dark fermentation bioreactors. *Microb. Cell Fact.* **2019**, *18*. [CrossRef]

54.	Yoshida, T.; Tashiro, Y.; Sonomoto, K. Novel high butanol production from lactic acid and pentose by *Clostridium Saccharoperbutylacetonicum*. *J. Biosci. Bioeng.* **2012**, *114*, 526–530. [CrossRef]

55.	Cabrol, L.; Marone, A.; Tapia-Venegas, E.; Steyer, J.-P.; Ruiz-Filippi, G.; Trably, E. Microbial ecology of fermentative hydrogen producing bioprocesses: Useful insights for driving the ecosystem function. *FEMS Microbiol. Rev.* **2017**, *41*, 158–181. [CrossRef]

56.	Altschul, S.F.; Gish, W.; Miller, W.; Myers, E.W.; Lipman, D.J. Basic local alignment search tool. *J. Mol. Biol.* **1990**, *215*, 403–410. [CrossRef]

57.	Zhao, X.; Condruz, S.; Chen, J.; Jolicoeur, M. A quantitative metabolomics study of high sodium response in *Clostridium acetobutylicum* ATCC 824 acetone-butanol-ethanol (ABE) fermentation. *Sci. Rep.* **2016**, *6*. [CrossRef] [PubMed]

58.	Patakova, P.; Branska, B.; Sedlar, K.; Vasylkivska, M.; Jureckova, K.; Kolek, J.; Koscova, P.; Provaznik, I. Acidogenesis, solventogenesis, metabolic stress response and life cycle changes in *Clostridium beijerinckii* NRRL B-598 at the transcriptomic level. *Sci. Rep.* **2019**, *9*. [CrossRef] [PubMed]

59.	Dou, T.-Y.; Chen, J.; Hao, Y.-F.; Qi, X. Effects of Different Carbon Sources on Enzyme Production and Ultrastructure of *Cellulosimicrobium cellulans*. *Curr. Microbiol.* **2019**, *76*, 355–360. [CrossRef] [PubMed]

60.	Wierzbicki, M.P.; Maloney, V.; Mizrachi, E.; Myburg, A.A. Xylan in the Middle: Understanding Xylan Biosynthesis and Its Metabolic Dependencies Toward Improving Wood Fiber for Industrial Processing. *Front. Plant. Sci.* **2019**, *10*. [CrossRef] [PubMed]

61.	Manfredi, P.; Renzi, F.; Mally, M.; Sauteur, L.; Schmaler, M.; Moes, S.; Jenö, P.; Cornelis, G.R. The genome and surface proteome of *Capnocytophaga canimorsus* reveal a key role of glycan foraging systems in host glycoprotein deglycosylation. *Mol. Microbiol.* **2011**, *81*, 1050–1060. [CrossRef]

62.	Kapke, P.A.; Brown, A.T.; Lillich, T.T. Carbon dioxide metabolism by *Capnocytophaga ochracea*: Identification, characterization, and regulation of a phosphoenolpyruvate carboxykinase. *Infect. Immun.* **1980**, *27*, 756–766.

 *fermentation*

Article

Bioethanol Production from Food Waste Applying the Multienzyme System Produced On-Site by *Fusarium oxysporum* F3 and Mixed Microbial Cultures

George Prasoulas [1], Aggelos Gentikis [1], Aikaterini Konti [2], Styliani Kalantzi [1], Dimitris Kekos [1] and Diomi Mamma [1,*]

1   Biotechnology Laboratory, School of Chemical Engineering, National Technical University of Athens, Zografou Campus, 15780 Athens, Greece; geo.prasoulas@gmail.com (G.P.); gentangelos@gmail.com (A.G.); stykalan@chemeng.ntua.gr (S.K.); kekos@chemeng.ntua.gr (D.K.)
2   Joint Research Centre (JRC), European Commission, 21027 Ispra, Italy; Aikaterini.KONTI@ec.europa.eu
*   Correspondence: dmamma@chemeng.ntua.gr; Tel.: +30-210-772-3160

Received: 27 February 2020; Accepted: 24 March 2020; Published: 26 March 2020

Abstract: Waste management and production of clean and affordable energy are two main challenges that our societies face. Food waste (FW), in particular, can be used as a feedstock for the production of ethanol because of its composition which is rich in cellulose, hemicellulose and starch. However, the cost of the necessary enzymes used to convert FW to ethanol remains an obstacle. The on-site production of the necessary enzymes could be a possible solution. In the present study, the multienzyme production by the fungus *Fusarium oxysporum* F3 under solid state cultivation using different agroindustrial residues was explored. Maximum amylase, glucoamylase, endoglucanase, b-glucosidase, cellobiohydrolase, xylanase, b-xylosidase and total cellulase titers on wheat bran (WB) were 17.8, 0.1, 65.2, 27.4, 3.5, 221.5, 0.7, 0.052 and 1.5 U/g WB respectively. *F. oxysporum* was used for the hydrolysis of FW and the subsequent ethanol production. To boost ethanol production, mixed *F. oxysporum* and *S. cerevisiae* cultures were also used. Bioethanol production by *F. oxysporum* monoculture reached 16.3 g/L (productivity 0.17 g/L/h), while that of the mixed culture was 20.6 g/L (productivity 1.0 g/L/h). Supplementation of the mixed culture with glucoamylase resulted in 30.3 g/L ethanol with a volumetric productivity of 1.4 g/L/h.

Keywords: bioethanol; food waste; on-site enzyme production; *Fusarium oxysporum* F3; *Saccharomyces cerevisiae*; mixed culture

1. Introduction

Food waste has been identified as a big economic, social and environmental problem nowadays. According to the official statistics published by Eurostat, each year, more than 240,000 t of waste is produced in the EU [1]. Bio-waste, the organic fraction of municipal solid waste, i.e., garden, kitchen and food waste, accounts for one third of the total waste and is considered to be a valuable resource that could be used as raw material for the production of high value-added products. This fact is also reflected in the updated Bioeconomy Strategy of the EU. A sustainable bioeconomy can turn bio-waste, residues and discards into valuable resources and can create innovations and incentives to help retailers and consumers cut food waste by 50% by 2030 [2]. However, the use of food by-products and the conversion of food waste is still limited. This is due to current limitations in its quantification along the food supply chain, limited data on its quality and level of homogeneity, and differences in national implementations of the waste legislation [2].

The composition of food waste, as already said, is not stable. It presents significant variations related to the season, the area, and the dietary habits of the population. Despite the inevitable variation in the composition of food waste, it can indisputably be said that it is rich in carbohydrates, proteins, lipids and minerals which make it an ideal raw material for the production of biofuels through microbial conversion [3,4]. The exploitation of food waste for the production of biofuels is also in line with the 2030 Agenda for Sustainable Development set by the UN in 2015 [5]. More precisely, it is directly related to the Sustainable Development Goals: 7. Affordable and Clean Energy, 12. Responsible Consumption and Production and 13. Climate Action. At the EU level, the importance of producing biofuels from bio-waste is reflected in the Renewable Energy Directive 2009/28/EC [6] and the recently adopted recast of the Renewable Energy Directive a.k.a. RED II [7]. The aforementioned legislation defines as advanced biofuels the 'biofuels that are produced from the feedstocks listed in Part A of Annex IX' that includes among others, the biomass fraction of municipal solid waste, the biomass fraction of industrial waste as well as bio-waste from households. The directive sets a sub-target of 3.5% for advanced biofuels within the 14% target for renewable energy in transport in 2030. Moreover, those biofuels will continue to count double towards the targets.

From a technical point of view, bioethanol production from lignocellulosic materials is a well-studied process and has been recently reviewed [8]. It includes the following processes: pretreatment, enzymatic hydrolysis, fermentation and ethanol recovery. The pretreatment phase aims at modifying the structural characteristics of the raw material facilitating the enzymes' access and maximizing sugar monomers production. The enzymatic hydrolysis targets the structural carbohydrates starch, cellulose and hemicellulose. During this step, pentoses and hexoses that can be further used in the fermentation step are liberated. In the subsequent fermentation step, microorganisms metabolize those readily available sugars, producing ethanol, which is subsequently recovered through distillation.

In terms of cost, the most demanding step, which significantly increases the total cost of the production of bioethanol and is identified as a barrier in the further deployment of ethanol production, is enzymatic hydrolysis [9]. A possible solution to this problem is the on-site production of the relevant enzymes instead of using commercially available enzymes [10]. Few organisms are able to produce the necessary enzymes; most of them belong to the species *Aspergillus sp., Penicillium sp., Trichoderma sp. Neurospora sp.* [11].

The filamentous fungus *Fusarium oxysporum* could be used both for cellulolytic enzyme production and for ethanol production because of its ability to ferment both hexoses and pentoses [12,13]. The main aim of the present work is, on the one hand, the induction of the metabolic system of *F. oxysporum* to produce the relevant enzymes by using different raw materials (namely wheat straw, wheat bran, corn cob), and on the other hand, to exploit those enzymes for the hydrolysis of food waste (FW) in order to produce ethanol. Focusing on increasing ethanol production, the addition of a glucoamylase was studied, as well as the use of mixed *F. oxysporum* and *Saccharomyces cerevisiae* cultures.

2. Materials and Methods

2.1. Raw Materials, Reagents, Microorganism

The food waste (FW) used in the present study was provided by Professor Gerasimos Lyberatos (Organic Chemical Technology Laboratory, School of Chemical Engineering, National Technical University of Athens), in dry form and with an average particle size of approximately 3 mm. FW were household food wastes from the Municipality of Halandri, Greece. The concept of drying focuses on the dehydration of the material resulting in a significant reduction of its mass and volume, thus facilitating its storage and protecting the readily fermentable sugars by inhibiting microbial activity due to low moisture content.

Wheat straw (WS), wheat bran (WB), and corn cobs (CC) were chopped into particles of less than 3 mm in diameter.

All chemicals were of analytical grade. Commercial glucoamylase (Spirizyme® Fuel) was kindly provided by Novozymes Corporation (Denmark).

The laboratory strain F3 of F. oxysporum, isolated from cumin [14], was used in the present study. The stock culture was maintained on potato-dextrose agar (PDA).

2.2. Chemical Analysis of FW

Moisture, ash, crude fat, crude protein (Kjeldhal method) and total starch content were determined according to standard methods [15]. Pectic polysaccharides were determined according to Phatak et al. [16], while water-soluble materials, cellulose, hemicellulose and acid insoluble lignin content as described by Sluiter et al. [17]. Analysis was carried out in triplicate.

2.3. Pretreatment of FW

Dry FW were pretreated as described elsewhere [18]. Briefly, FW at concentration of 30% w/v was pretreated at 100 °C for 1 h in the presence of 1 g sulfuric acid/100 g of dry FW. Following pH adjustment to 6.0, the pretreated material (the whole slurry) was used in fermentation experiments.

2.4. Media and Growth Conditions for the in Situ Production of Enzymes

Solid-state cultivation (SSC) was carried out in 100-mL Erlenmeyer flasks containing 3.0 g of dry carbon source (WB, WS or CC) moistened with Toyama's mineral medium (in $g \cdot L^{-1}$: $(NH_4)_2SO_4$, 10; KH_2PO_4, 3; $MgSO_4 \cdot 7H_2O$, 0.5; $CaCl_2$, 0.5) [19]. The initial culture pH was adjusted to 6.0 and the moisture level at 75%. Following heat sterilization (121°C) for 20 min, each flask was inoculated with 1 mL spore suspension (approximately 10^7 conidia) and incubated at 30 °C under static conditions. Experiments were carried out in duplicate.

2.5. Enzyme Extraction

After suitable periods of time, enzymes were extracted from the SSC with 10-fold (v/w) 50 mM citrate-phosphate buffer pH 6.0 by shaking (250 rpm) at 25 °C for 60 min. The suspended materials and fungal biomass were separated by centrifugation (12,000 x g at 4 °C for 15 min) and the clarified supernatant was used for enzyme activity measurements.

2.6. Enzymatic Hydrolysis of FW

SSC at maximum enzyme production was supplemented with the pretreated FW (SSC to FW ratio, 1/10 w/w) and the appropriate amount of the commercial glucoamylase (Spirizyme® Fuel) in order to achieve 20 and 40 Units of glucoamylase per g starch. Microbial contamination was prevented by the addition of sodium azide (0.02% w/v). Hydrolysis was performed at 50 ± 1 °C in a rotary shaker (250 rpm). Samples were withdrawn periodically, centrifuged (10,000 x g for 10 min), and analyzed for glucose.

2.7. Conversion of FW into Bioethanol

Food waste was converted into ethanol applying of a two-phases process where enzymes production under SSC was combined with simultaneous saccharification and fermentation (SSF) of FW by the mesophilic fungus F. oxysporum F3 or by a mixed culture of the latter with the yeast S. cerevisiae. Initially F. oxysporum F3 was grown under SSC, as described above for the production of the cellulolytic, hemicellulolytic and amylolytic enzymes. Whole SSC (fungal mycelia and the in situ produced enzymes) was transferred to the pretreated FW (SSC to FW ratio, 1/10 w/w). Fermentation was carried out in a rotary shaker operating at 30 ± 1 °C and 80 rpm in Erlenmeyer flasks provided with special rubber stoppers, which ensured anaerobic conditions and allowed release of produced carbon dioxide.

In the case of mixed microbial culture compressed baker's yeast (Yiotis, Athens, Greece) corresponding to 15 mg yeast per gram of initial dry FW was added.

2.8. Analytical Methods

Endoglucanase (EG), exoglucanase (EXG), xylanase (XYL), total cellulase (FPU) and amylase (AMYL) activities were assayed on carboxymethyl cellulose, Avicel, birchwood xylan filter paper and starch respectively, as described [18,20]. The activities of β-glucosidase (β-GLU) β-xylosidase (β-XYL) and glucoamylase (GLAMYL) were determined spectrophotometrically using the respective p-nitrophenyl glycosides as substrates [18,20]. All assays were carried out at 50 °C and pH 5.0. Blanks with inactivated enzyme (after boiling for 15 min) were used as a reference.

One unit (U) of enzyme activity was defined as the amount of enzyme liberating 1 µmole of product per min.

Glucose concentration was determined using a commercially available kit (Biosis S.A., Athens, Greece) that employed the Glucose Oxidase–Peroxidase (GOX–PER) method.

Ethanol analysis was conducted in an Aminex HPX-87H (Bio-Rad, 300 x 7.8 mm, particle size 9 µm) chromatography column. Mobile phase was 5 mM H_2SO_4 in degassed HPLC- water at a flow rate of 0.6 mL/min and column temperature was 40 °C [21].

3. Results and Discussion

3.1. Food Waste Composition

In general, the composition of food waste is complex and includes oil and water, as well as spoiled and leftover foods from kitchen wastes and markets. These substances are mainly composed of soluble sugars, carbohydrate polymers (pectin, starch, cellulose and hemicelluloses), lignin, proteins, lipids and a remaining, smaller inorganic part. High moisture content leads to rapid decomposition of the organic wastes and the production of unpleasant odors, which can attract flies and bugs, which are vectors for various diseases [22]. The feedstock used in the present study was provided in dry form as mentioned above. Compositional analysis of FW is presented in Table 1.

Table 1. Chemical analysis of dried FW.

Component	%(w/w, Dry Basis)
Water Soluble Materials [a]	27.29 ± 1.71
Starch	10.68 ± 0.07
Cellulose	10.31 ± 0.07
Hemicellulose	11.32 ± 0.02
Pectin	3.27 ± 0.82
Protein	13.70 ± 3.31
Lipids	12.26 ± 0.11
Lignin	6.75 ± 0.16
Ash	5.16 ± 0.20

[a] Total reducing sugars: 4.96 ± 0.94 % (w/w, dry basis), sucrose: 0.51 ± 0.04 % (w/w, dry basis), protein: 0.33 ± 0.02 (w/w, dry basis) and soluble starch: 1.15 ± 0.09 (w/w, dry basis).

The composition of FW, carbohydrates, protein, makes it an excellent feedstock for the production of biofuels and bio-based chemicals through microbial conversion. The polysaccharide content of FW is difficult to be used by ethanol producing microorganisms such as Saccharomyces cerevisiae. Different commercial enzymes (amylase, glucoamylase, carbohydrase, cellulase) have been used to improve the saccharification of FW [23].

3.2. Multienzyme Production under Solid-State Cultivation

To make the enzymatic hydrolysis of FW more cost-effective, the enzymes should be produced on-site from a cheap feedstock [11]. SSC has several biotechnological advantages such as higher fermentation capacity, higher end-product stability, lower catabolic repression and cost-effective technology [24,25]. SSC is an attractive process for filamentous fungus cultivation because the solid substrates have characteristics similar to the natural habitat of the fungi, resulting in improved growth and secretion of a wide range of enzymes. Selecting the appropriate substrate is an extremely important aspect of SSC, as the solid material will act as a physical support and nutrient source [26]. In the present study three different agroindustrial residues, wheat straw (WS), wheat bran (WB) and corn cobs (CC) were evaluated for the production of cellulolytic, hemicellulolytic and amylolytic enzymes by the mesophilic fungus *F. oxysporum* F3. Wheat bran (WB), a nutrient-richer intermediate of the wheat processing industry, was the most effective carbon source for multi-enzyme production by *F. oxysporum* F3 (Figure 1). Implementation of corn cobs (CC) and wheat straw (WS) particles as substrates was associated with lower enzyme titres. Maximum enzyme activities were recorded in the 5th day of fermentation. Amylase activity was found 5- fold higher, than that produced on WS and CC, while traces of glucoamylase was produced on WS and CC. Maximum endoglucanase, b-glucosidase cellobiohydrolase, xylanase and b-xylosidase titers on WB were 65.2, 27.4, 3.5, 221.5, 0.7, 0.052 U/g WB, respectively, while total cellulase activity was 1.5 FPU/g WB.

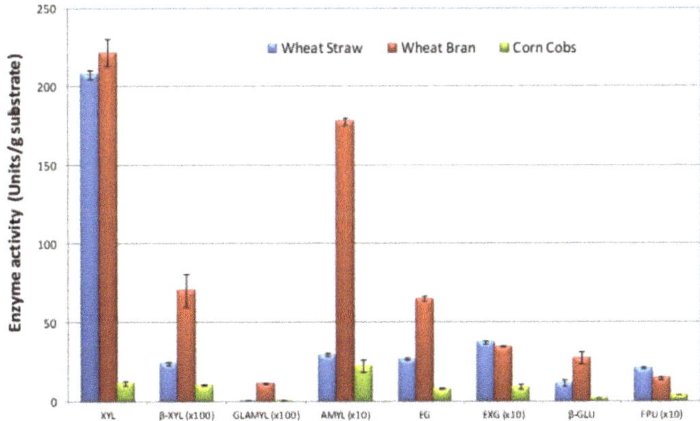

Figure 1. Production of cellulolytic, hemicellulolytic and amylolytic enzymes by *F. oxysporum* F3 grown under solid state cultivation on wheat straw, wheat bran and corn cobs as carbon sources.

For comparison, *F. oxysporum* F3 grown under SSC on corn stover as carbon source resulted in final endoglucanase, b-glucosidase, cellobiohydrolase, xylanase, b-xylosidase titers of 211, 0.088, 3.9, 1216, 0.052 U/g, respectively [27]. Futhermore, *Trichoderma virens* grown on alkali-treated WS under SSC produced 123.26 and 348 U/g endoglucase and xylanase, respectively [28].

It is well documented that the type and composition of the carbohydrates present in WB are suitable for induction of cellulases, hemicellulases and amylases from filamentous fungi under SSC [24,25]. Since the target of the present study is the production of a multi-enzyme system capable of hydrolysing the main polysaccharides present in FW, WB was chosen for further experiments.

3.3. Hydrolysis of FW by the On-Site Produced Multienzyme System of F. Oxysporum F3

Multienzyme system of F. oxyporum F3 produced under SSC on WB as carbon source was evaluated in FW hydrolysis. Furthermore, due to low glucoamylase titer produced the hydrolysis mixture was supplemented with 20 and 40 Units/g FW. The concentration of glucose increased gradually with time and reached a constant value at 69 h (Figure 2a). Glucose release applying F. oxysporum enzyme system was found 33.7 g/L corresponding to a hydrolysis yield of 47.4% (based on cellulose and starch content of FW). It should be mentioned that glucose release due to pretreatment of FW accounted for 11.2 g/L. Supplementation of hydrolysis mixture with glucoamylase (Spirizyme® Fuel) increased glucose release by approximately 25%.

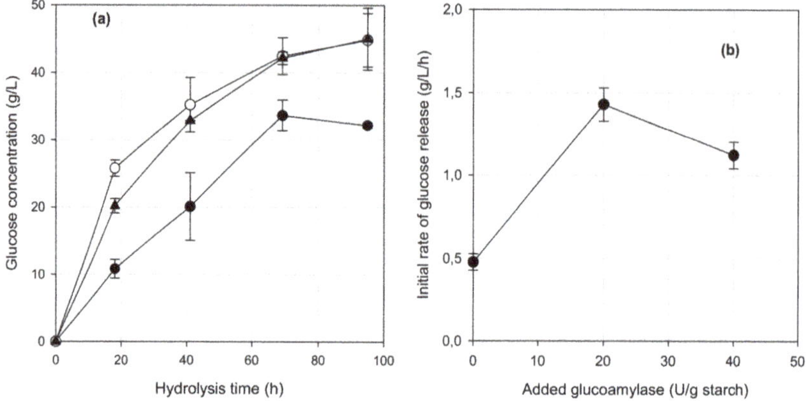

Figure 2. (**a**) Time course of glucose release during FW hydrolysis by *F. oxysporum* F3 multienzyme system (•), supplemented with 20 (○), 40 (▲) Units/g starch Spirizyme® Fuel. (**b**) Initial rate of glucose release.

The amount of added glucoamylase affects the initial rate of glucose release (Figure 2b) during hydrolysis. Addition of 20 U/g starch glucoamylase resulted in a 66% improvement in the rate of glucose release while further increase in glucoamylase load did not improve the result.

3.4. Bioethanol Production by Mixed Microbial Culture of F. Oxysporum F3 with the Yeast S. Cerevisiae

The processes generally used in bioethanol production are simultaneous saccharification and fermentation (SSF) and separate hydrolysis and fermentation (SHF). Performing hydrolysis and fermentation in a single step, the SSF process, has several advantages over SHF. In SSF, end-product inhibition of b-glucosidase is avoided, and the need for separate reactors is eliminated [29]. In the present study FW were converted to bioethanol applying SSF process in batch mode. Ethanol production from FW by mono-culture of *F. oxysporum* F3 reached 16.3 g/L after 94 h of fermentation (Figure 3a) corresponding to 31.8% of the maximum theoretical based on the soluble fraction and the carbohydrate content (cellulose, starch and hemicellulose) of FW (Figure 4).

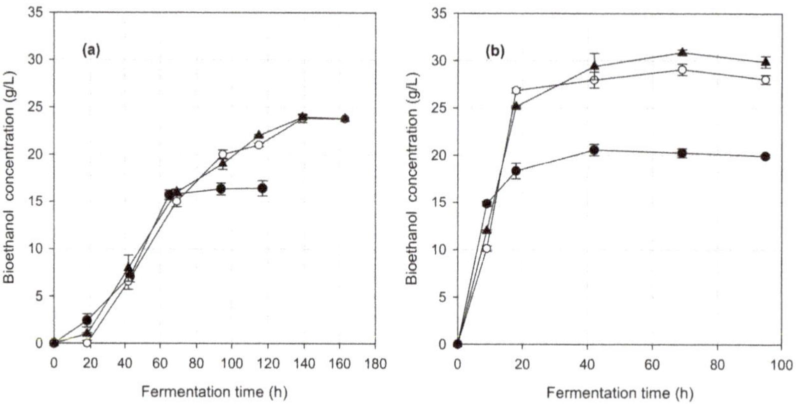

Figure 3. Bioethanol production from FW by (**a**) mono-culture of F. oxysporum F3 (•), supplemented with 20 (○) and 40 (▲) Units/g starch Spirizyme® Fuel; and (**b**) mixed microbial culture of F. oxysporum F3 with the yeast S. cerevisiae (•), supplemented with 20 (○) and 40 (▲) Units/g starch Spirizyme® Fuel.

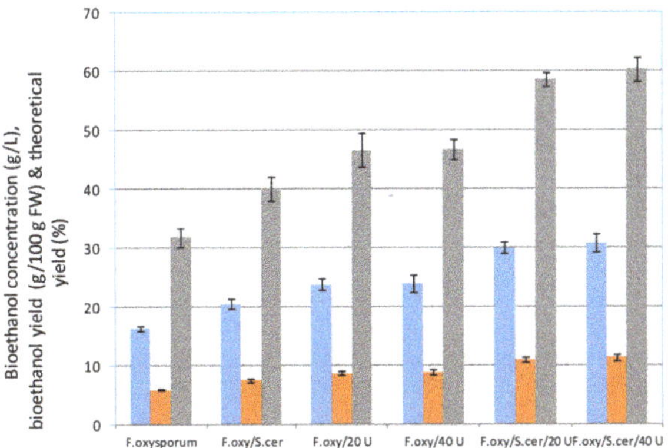

Figure 4. Maximum values of bioethanol production (g/L) (■), bioethanol yield (g/100 g FW) (■), theoretical yield (%) (■).

When the fermentation medium was supplemented with 20 U/g starch glucoamylase, ethanol titer was 23.9 g/L (corresponding to 46.8.0% of the maximum theoretical). Increase in glucoamylase supplementation did not improve bioethanol production (Figure 3a). Maximum ethanol production was recorded at 139 h of fermentation, and no further increase in ethanol concentration was found when fermentation was extended to 163 h. Bioethanol volumetric productivities were in the range of 0.17 to 0.23 g/L/h (Figure 5).

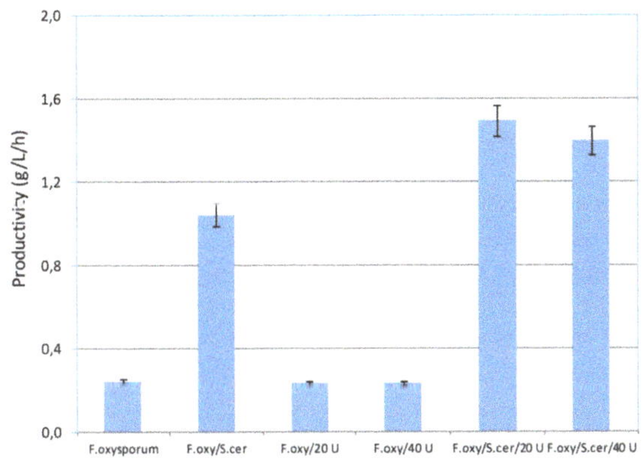

Figure 5. Bioethanol volumetric productivities.

To improve the sugar assimilation rate, mixed culture of *F. oxysporum* F3 with the yeast *S. cerevisiae* were applied with the same bioconversion setup. The results are presented in Figure 3b. As can be seen, ethanol concentration at 18 h of process was about 18.3 g/L in the mixed culture of *F. oxysporum* F3 with *S. cerevisiae* approximately 7.5 times higher than the corresponding value achieved with the mono-culture of *F. oxysporum* F3. Fermentation was completed after 42 h using mixed microbial culture and ethanol production reached 20.6 g/L (corresponding to 40.1% of the maximum theoretical) (Figure 4). Ethanol concentrations of about 26.8 and 25.1 g/L were reached at 18 h when the mixed cultures were supplemented with 20 and 40 U/g starch of glucoamylase, while the corresponding values in mono-cultures were in the range of 1 g/L. Maximum ethanol production of 29.9 and 30.8 g/L was achieved at 69 h (Figure 3b). At the end of all fermentations, glucose concentration in the fermentation broth was lower than 2.0 g/L. Volumetric productivities ranged from 1.0 g/L/h (mixed microbial culture) to 1.5 g/L/h (mixed microbial culture supplemented with 20 U/g glucoamylase) (Figure 5).

It is evident that mixed microbial culture decreases significantly the time needed for the fermentation to be completed therefore increasing volumetric productivity. Mixed cultures of mesophilic fungi, such as *F. oxysporum* F3 and *N. crassa*, with the yeast *S. cerevisiae*, have been successfully implemented for the bioconversion of sweet sorghum and sweet sorghum baggasse to bioethanol [30,31]. Table 2 summarizes results obtained from the bioconversion of different kinds of FW.

Table 2. Production of bioethanol from FW applying different processes.

Microorganism	Type of Process	Enzymes Used	Ethanol (g/L)	Productivity (g/L/h)	Reference
S.cerevisiae	SHF	amylase, glucoamylase	8.0	0.33	[32]
S. cerevisiae (dry baker's yeast)	SHF	on-site produced enzymes [1]	19.7	0.92	[33]
S. cerevisiae	SSF	Carbohydrase [2]	20.0	0.8	[34]
S. cerevisiae (dry baker's yeast)	SHF	amylase glucoamylase cellulase b-glucosidase	23.3	0.49	[35]
S. cerevisiae	SSF	glucoamylase	33.0	0.49	[36]
S. cerevisiae (dry baker's yeast)	SHF	on-site produced enzymes [2]	58.0	1.8	[37]
F. oxysporum- S. cerevisiae	SSF	on-site produced enzymes + glucoamylase	30.8	1.4	Present study

[1] Cellulolytic enzymes produced from the thermophillic fungus Myceliophthora thermophila, [2] cellulolytic enzyme system [3] amylolytic, cellulolytic and hemicellulolytic enzymes produced by Aspergillus awamori.

Bioethanol production of the present sudy was 30.8 g/L which is higher than that reported by Matsakas and Christakopoulos [33] using the on-site produced cellulolytic enzymes from the thermophillic fungus Myceliophthora thermophila. On the other hand, Kiran and Liu [37] achieved much higher ethanol production (58.0 g/L) using the multienzyme system produced on-site by the fugus Aspergillus awamori. Wang et al. [36] used response surface methodology to optimize the conditions of SSF for ethanol production from kitchen garbage. Maximum ethanol concentration of 33.0 g/L was reported with the optimum conditions of time of 67.60 h, pH= 4.18 and T = 35 °C using glucoamylase in the saccharification step. Kitchen wastes were treated with a mixture of amylolytic and cellulolytic enzymes, the hydrolyzate was converted to bioethanol using commercial dry baker's yeast, and resulted in 23.3 g/L bioethanol production [35].

Bioethanol volumetric productivities ranged from 0.33 to 1.8 g/L/h and that of the present study is among the higher (Table 2). It is evident that the variability in FW content from region to region, the substrate concentration, the type and dosage of used enzymes and processes applied affect bioethanol production.

Enzyme cost still remains high, and this is identified as a major challenge in the deployment of lignocellulosic ethanol [9]. If the necessary enzymes could be efficiently produced on-site, the cost could be significantly reduced. A recent study has estimated that this cellulase cost can be reduced, from 0.78 to 0.58 $/gallon, by shifting from the off-site to the on-site approach of cellulase production [38].

4. Conclusions

In the present study, the feasibility of producing bioethanol from FW applying the on-site produced multienzyme system was demonstrated. The mesophilic fungus *F. oxysporum* F3 grown under SSC in wheat bran as carbon source produced a mixture of hydrolytic enzymes capable of hydrolyzing the polysaccharides in FW. Mixed-microbial cultures of *F. oxysporum* with the yeast *S. cerevisiae* increased bioethanol volumetric productivity compared to mono-culture of the fungus. Bioethanol production increased by approximately 23% when the mixed microbial culture was supplemented with commercial glucoamylase. The results of the study demonstrated that non-commercial enzyme products obtained from fungi could be an efficient alternative to commercial preparations in technologies which use elevated substrate loadings or where an accurate loading is impossible due to practical limitations.

Author Contributions: Conceptualization, A.K., D.K. and D.M.; Methodology, A.K. and D.M.; Investigation, G.P., A.G., S.K.; Writing—Original Draft Preparation, D.M.; Writing—Review & Editing, all authors. All authors have read and agreed to the published version of the manuscript.

Funding: This research received no external funding

Acknowledgments: The authors would like to thank Gerasimos Lyberatos of Organic Chemical Technology Laboratory, School of Chemical Engineering, National Technical University of Athens for providing the dry FW. The authors would also like to thank Novozymes Corporation for generously providing the cellulose enzyme samples.

Conflicts of Interest: The authors declare no conflict of interest.

References

1. Eurostat. *Municipal Waste by Waste Operations*; Eurostat: Brussels, Belgium, 2017.
2. European Commission. *A sustainable Bioeconomy for Europe: Strengthening the Connection between Economy, Society and the Environment—Updated Bioeconomy Strategy*; European Commission: Brussels, Belgium, 2018.
3. Patel, A.; Hrůzová, K.; Rova, U.; Christakopoulos, P.; Matsakas, L. Sustainable biorefinery concept for biofuel production through holistic volarization of food waste. *Bioresour. Technol.* **2019**, *294*, 122247. [CrossRef] [PubMed]
4. Carmona-Cabello, M.; García, I.L.; Sáez-Bastante, J.; Pinzi, S.; Koutinas, A.A.; Dorado, M.P. Food waste from restaurant sector—Characterization for biorefinery approach. *Bioresour. Technol.* **2020**, *301*, 122779. [CrossRef] [PubMed]
5. UN. *Transforming Our World: The 2030 Agenda for Sustainable Development, A/Res/70/1*; United Nations: New York, NY, USA, 2015.
6. European Parliament and Council of the European Union. *Directive 2009/28/EC of the European Parliament and of the Council of 23 April 2009 on the Promotion of the Use of Energy from Renewable Sources and Amending and Subsequently Repealing Directives 2001/77/EC and 2003/30/EC*; European Parliament and Council of the European Union: Brussels, Belgium, 2009.
7. European Parliament and Council of the European Union. *Directive 2018/2001 of the European Parliament and of the Council of 11 December 2018 on the Promotion of the Use of Energy from Renewable Sources (Recast)*; European Parliament and Council of the European Union: Brussels, Belgium, 2018.
8. Barampouti, E.M.; Mai, S.; Malamis, D.; Moustakas, K.; Loizidou, M. Liquid biofuels from the organic fraction of municipal solid waste: A review. *Renew. Sustain. Energy Rev.* **2019**, *110*, 298–314. [CrossRef]
9. Padella, M.; O'Connell, A.; Prussi, M. What is still limiting the deployment of cellulosic ethanol? Analysis of the Current Status of the Sector. *Appl. Sci.* **2019**, *9*, 4523. [CrossRef]
10. Wan, Y.K.; Sadhukhan, J.; Ng, D.K.S. Techno-economic evaluations for feasibility of sago-based biorefinery, Part 2: Integrated bioethanol production and energy systems. *Chem. Eng. Res. Des.* **2016**, *107*, 102–116. [CrossRef]
11. Siqueira, J.G.W.; Rodrigues, C.; Vandenberghe, L.P.D.S.; Woiciechowski, A.L.; Soccol, C.R. Current advances in on-site cellulase production and application on lignocellulosic biomass conversion to biofuels: A review. *Biomass Bioenerg* **2020**, *132*, 105419. [CrossRef]
12. Da Rosa-Garzon, N.G.; Laure, H.J.; Rosa, J.C.; Cabral, H. Fusarium oxysporum cultured with complex nitrogen sources can degrade agricultural residues: Evidence from analysis of secreted enzymes and intracellular proteome. *Renew. Energ.* **2019**, *133*, 941–950. [CrossRef]
13. Sharma, S.; Kuila, A.; Sharma, V. Enzymatic hydrolysis of thermochemically pretreated biomass using a mixture of cellulolytic enzymes produced from different fungal sources. *Clean Technol. Environ.* **2017**, *19*, 1577–1584. [CrossRef]
14. Christakopoulos, P.; Macris, B.J.; Kekos, D. Direct fermentation of cellulose to ethanol by Fusarium oxysporum. *Enzyme Microb. Technol.* **1989**, *11*, 236–239. [CrossRef]
15. AOAC. *Official Methods of Analysis of the Association of Official Analytical Chemists*, 15th ed.; AOAC Inc.: Rockville, MD, USA, 1990.

16. Phatak, L.; Chang, K.C.; Brown, G. Isolation and characterization of pectin in sugar-beet pulp. *J. Food Sci.* **1988**, *53*, 830–833. [CrossRef]

17. Sluiter, A.; Hames, B.; Ruiz, R.; Scarlata, C.; Sluiter, J.; Templeton, D.; Crocker, D. *Determination of Structural Carbohydrates and Lignin in Biomass*; Technical Report NREL/TP-510-42618; National Renewable Energy Laboratory: Golden, CO, USA, 2012. Available online: https://www.nrel.gov/docs/gen/fy13/42618.pdf (accessed on 8 January 2020).

18. Loizidou, M.; Alamanou, D.G.; Sotiropoulos, A.; Lytras, C.; Mamma, D.; Malamis, D.; Kekos, D. Pilot scale system of two horizontal rotating bioreactors for bioethanol production from household food waste at high solid concentrations. *Waste Biomass Valorization* **2017**, *8*, 1709–1719. [CrossRef]

19. Toyama, N.; Ogawa, K. Cellulase production by *Trichoderma viride* in solid and submerged culture conditions. In Proceedings of the Symposium for the Bioconversion of Cellulosic Substances into Energy, Chemicals and Microbial Protein, New Delhi, India, 21–23 February 1977; Ghose, T.K., Ed.; IIT: New Delhi, India, 1978; pp. 325–327.

20. Dogaris, I.; Vakontios, G.; Kalogeris, E.; Mamma, D.; Kekos, D. Induction of cellulases and hemicellulases from *Neurospora crassa* under solid-state cultivation for bioconversion of sorghum bagasse into ethanol. *Ind. Crop. Prod.* **2009**, *29*, 404–411. [CrossRef]

21. Alamanou, D.G.; Malamis, D.; Mamma, D.; Kekos, D. Bioethanol from dried household food waste applying non-isothermal simultaneous saccharification and fermentation at high substrate concentration. *Waste Biomass Valorization* **2015**, *6*, 353–361. [CrossRef]

22. Yazid, N.A.; Barrena, R.; Komilis, D.; Sánchez, A. Solid-State Fermentation as a Novel Paradigm for Organic Waste Valorization: A Review. *Sustainability* **2017**, *9*, 224. [CrossRef]

23. Kiran, E.U.; Trzcinski, A.P.; Ng, W.J.; Liu, Y. Bioconversion of food waste to energy: A review. *Fuel* **2014**, *134*, 389–399. [CrossRef]

24. Behera, S.S.; Ray, R.C. Solid state fermentation for production of microbial cellulases: Recent advances and improvement strategies. *Int. J. Biol. Macromol.* **2016**, *86*, 656–669. [CrossRef]

25. Sadh, P.K.; Duhan, S.; Duhan, J.S. Agro-industrial wastes and their utilization using solid state fermentation: A review. *Bioresour. Bioprocess.* **2018**, *5*, 1. [CrossRef]

26. De Castro, R.J.S.; Sato, H.H. Enzyme Production by Solid State Fermentation: General Aspects and an Analysis of the Physicochemical Characteristics of Substrates for Agro-industrial Wastes Valorization. *Waste Biomass Valorization* **2015**, *6*, 1085–1093. [CrossRef]

27. Panagiotou, G.; Kekos, D.; Macris, B.J.; Christakopoulos, P. Production of cellulolytic and xylanolytic enzymes by Fusarium oxysporum grown on corn stover in solid state fermentation. *Ind. Crop. Prod.* **2003**, *18*, 37–45. [CrossRef]

28. El-Shishtawy, R.M.; Mohamed, S.A.; Asiri, A.M.; Gomaa, A.A.; Ibrahim, I.H.; Al-Talhi, H.A. Saccharification and hydrolytic enzyme production of alkali pre-treated wheat bran by Trichoderma virens under solid state fermentation. *BMC Biotechnol.* **2015**, *15*, 37. [CrossRef]

29. Gupta, A.; Verma, J.P. Sustainable bio-ethanol production from agro-residues: A review. *Renew. Sustain. Energy Rev.* **2015**, *41*, 550–567. [CrossRef]

30. Mamma, D.; Koullas, D.; Fountoukidis, G.; Kekos, D.; Macris, B.J.; Koukios, E. Bioethanol from Sweet Sorghum: Simultaneous saccharification and fermentation of carbohydrates by a mixed microbial culture. *Process Biochem.* **1996**, *31*, 377–381. [CrossRef]

31. Dogaris, I.; Gkounta, O.; Mamma, D.; Kekos, D. Bioconversion of dilute-acid pretreated sorghum bagasse to ethanol by *Neurospora crassa*. *Appl. Microbiol. Biot.* **2012**, *95*, 541–550. [CrossRef] [PubMed]

32. Walker, K.; Vadlani, P.; Madl, R.; Ugorowski, P.; Hohn, K.L. Ethanol Fermentation from Food Processing Waste. *Environ. Prog. Sustain.* **2013**, *32*, 1280–1283. [CrossRef]

33. Matsakas, L.; Christakopoulos, P. Ethanol Production from Enzymatically Treated Dried Food Waste Using Enzymes Produced On-Site. *Sustainability* **2015**, *7*, 1446–1458. [CrossRef]

34. Kim, J.H.; Lee, J.C.; Pak, D. Feasibility of producing ethanol from food waste. *Waste Manag.* **2011**, *31*, 2121–2125. [CrossRef]

35. Cekmecelioglu, D.; Uncu, O.N. Kinetic modeling of enzymatic hydrolysis of pretreated kitchen wastes for enhancing bioethanol production. *Waste Manag.* **2013**, *33*, 735–739. [CrossRef]

36. Wang, Q.; Ma, H.; Xu, W.; Gong, L.; Zhang, W.; Zou, D. Ethanol production from kitchen garbage using response surface methodology. *Biochem. Eng. J.* **2008**, *39*, 604–610. [CrossRef]

37. Kiran, E.U.; Liu, Y. Bioethanol production from mixed food waste by an effective enzymatic pretreatment. *Fuel* **2015**, *159*, 463–469. [CrossRef]
38. Johnson, E. Integrated enzyme production lowers the cost of cellulosic ethanol. *Biofuels Bioprod. Biorefin.* **2016**, *10*, 164–174. [CrossRef]

 *fermentation*

Article

Biodegradation of Residues from the Palo Santo (*Bursera graveolens*) Essential Oil Extraction and Their Potential for Enzyme Production Using Native *Xylaria* Fungi from Southern Ecuador

Vinicio Carrión-Paladines [1,*], Andreas Fries [2], Rosa Elena Caballero [3], Pablo Pérez Daniëls [4] and Roberto García-Ruiz [5]

[1] Departamento de Ciencias Biológicas, Sección de Ecología y Sistemática, Universidad Técnica Particular de Loja, San Cayetano Alto s/n C.P., Loja 110160, Ecuador
[2] Departamento de Geología, Minas e Ingeniería Civil (DGMIC), Universidad Técnica Particular de Loja, San Cayetano Alto s/n C.P., Loja 110160, Ecuador
[3] Departamento de Química, Universidad Autónoma de Chiriquí, El Cabrero, David, Chiriquí PO Box 0427, Panama
[4] Departamento de Botánica, Ecología y Fisiología Vegetal, Universidad de Córdoba, Ed. Celestino Mutis, 3ª pta., Campus de Rabanales, 14071 Córdoba, Spain
[5] Departamento de Biología Animal, Biología Vegetal y Ecología, Sección de Ecología, Universidad de Jaén, España, Campus Las Lagunillas, Edificio Ciencias Experimentales y de la Salud (B3), 23071 Jaén, Spain
* Correspondence: hvcarrionx@utpl.edu.ec; Tel.: +593-7-2570-275 (ext. 3034)

Received: 9 July 2019; Accepted: 17 August 2019; Published: 23 August 2019

Abstract: The degradation dynamics of lignin and cellulose were analyzed by means of a solid state biodegradation experiment, using residues from the essential oil extraction of the Palo Santo tree (*Bursera graveolens*). As such, two native *Xylaria* spp. and an exotic mushroom *Trametes versicolor* were incubated on the spent substrate (Residues of *B. Graveolens*, BGR's). The relatively high lignin and cellulose contents of the BGRs (9.1% and 19%, respectively) indicated the potential of this resource for the production of methane (biogas) and ethanol. However, the degradation of the lignin and cellulose content could be traced back to the relatively high activity of the enzymes laccase, cellulase, and xylanase, produced by the fungi. The results showed that laccase (30.0 U/L and 26.6 U/L), cellulase (27.3 U/L and 35.8 U/L) and xylanase (189.7U/L and 128.3 U/L) activities of *Xylaria feejeensis* and *Xylaria* cf. *microceras* were generally higher than *T. versicolor* (9.0 U/L, 29.5 U/L, 99.5 U/L respectively). Furthermore, the total carbon (TC: 47.3%), total nitrogen (TN: 1.5%), total phosphorus (TP: 0.2%) and total potassium (TK: 1.2%) dynamics were analyzed during the experiment and their importance for the degradation process highlighted. The results of this work might serve as guidance for future studies in dry forest areas, while furthering the understanding of the potential use of native fungi as ecologic lignocellulosic decomposers and for industrial proposes.

Keywords: Decomposition dynamics; *Bursera graveolens* waste; Xylariaceae; laccase; cellulase; xylanase

---

1. Introduction

The Palo Santo (*Bursera graveolens* [Kunth] Triana & Planchon; Burseraceae) is a native, deciduous, dioecious, non-timber tree species of dry forest areas and is distributed from western Mexico to northwestern Peru. In Ecuador, *B. graveolens* is native in the western coastal plains and also on the Galapagos Islands [1]. Particularly, in the dry forest areas of southern Ecuador, the Palo Santo tree is the most dominant native tree species [2].

The Palo Santo is considered a vital resource for the local communities of the dry forest, as different parts of the tree are used in traditional medicine, as well as for the extraction of essential oil [3]. In addition, the wood and stalks are applied to prevent mosquito bites and to treat aches and pains of differing origins, such as fibrosarcoma, atherosclerosis, and arthritis [4]. The production of essential oil from different parts of the Palo Santo tree has grown during the last few decades due to the increase of the global demand within the cosmetic and pharmacological industry. Generally, the woody material of the tree is used for essential oil extraction. However, currently it is also extracted from the fruits, as is practiced in Ecuador.

The essential oil extraction process generates abundant organic waste, which is commonly discarded directly into the natural ecosystems or burned. The organic waste can cause environmental problems such as air pollution and/or water and soil contamination because of the low natural degradation capacity of these residues. The essential oil content of the different parts of *B. graveolens* is relatively high. The wood in the form of kindling contains up to 5.2% of essential oil and the shavings up to 3.4% [3]. Using the fruits, the distillation process is less efficient because the fresh fruits only contain up to 3% of essential oil. Therefore, a considerable amount of organic waste is concurrently produced (at least 95% of the fresh biomass), for which reason waste management is necessary.

One possible reuse of this waste is the production of vermicomposts for farming purposes. However, the waste must be mixed with other organic residues, like kitchen waste or animal manures, to make the product suitable for agricultural proposes [5]. Another reuse potential is enzyme production for industrial purposes by means of biodegradation of the wastes with specific fungi. Indeed, microbial enzymes are known to play a crucial role as metabolic catalysts, which is why they are frequently used in various industries among other applications. The use of enzymes is extremely wide-spread, especially in industries [6] where over 500 products are made of enzymes and about 150 industrial processes need enzymes as microbial cell catalysts [7]. Therefore, the demand for industrial enzymes is continuously rising, which leads to a growing need for sustainable solutions. Microbes are one of the largest and most useful sources of enzymes in nature [8], but research is still needed due to the immense biodiversity, particularly in tropical countries. Analyzing the diversity of potential substrates and microorganisms in different ecosystems, such as the dry forest of Ecuador, may expand the knowledge about enzyme production sources.

Some of the enzymes which are frequently used for industrial application are cellullase, laccase, and xylanase. These enzymes detoxify industrial effluents from the paper, pulp, textile, and petrochemical industries. Furthermore, they are used in medical diagnostic tools, as catalysts in drug manufacturing, and as cleaning agents for water purification systems. Additionally, these enzymes are needed as ingredients in the cosmetic industry, as well as for the bioremediation of herbicides and pesticides [9].

Cellulase is especially important as a detergent additive because it catalyzes the breakdown of chemical bonds and is used in the textile industry for cleaning processes and to reduce waste production. Besides this, cellulase contributes to the sustainable production of second generation biofuels and other chemical derivatives [10]. Laccase is used for decoloration of textile effluents and textile bleaching [11], as well as to oxidize phenolic and non-phenolic lignin-related compounds and other environmental pollutants [12]. Moreover, laccases are also used in the formulation of biofuels, biosensors and the synthesis of new hybrid molecules [13]. Xylanase is mainly needed in the pulp, paper, food, and beverage industries, as well as for the saccharification of pre-treated lignocellulosic biomass for the production of second generation biofuels [14].

In the tropics, a prominent fungus genus used for biodegradation and enzyme production is *Xylaria* Hill ex Schrank (Xylariaceae, Xylariales, Sordariomycetes, Ascomycota), which comprises more than 300 species [15]. *Xylaria* spp. are considered to be significant producers of different ligninolytic enzymes, including laccase, xylanase and cellulase [16], besides their ability to degrade lignin [17], hemicellulose [18], and cellulose from trees and agricultural residues [19–22]. Within the ecosystems, this type of fungus participates in the carbon and nitrogen cycles [23] and plays an important role in the biodegradation processes of wood and leaf litter due to its complex and diverse enzymatic system.

In other biotechnological applications, the spent substrates are analyzed, especially the amounts of lignin, hemicellulose and cellulose. Lignin and hemicellulose are used in bioconversion processes for the production of bioethanol, biogas, and other biofuel products, while cellulose is used in the manufacture of cosmetics as well as in the development of new renewable energy sources [24]. Besides these parameters, the C/N ratio of the spent substrate is important, because a high C/N ratio may cause the immobilization of nutrients, which limits some biological processes such as respiration rates and the development of microbial biomass [25]. Furthermore, inorganic nutrients (phosphorus and potassium) of spent substrates are essential during the biodegradation process, because these elements can restrict the degradation of the spent substrate and the development of the microbial cells [26].

As far as we know, residues of *B. graveolens* obtained during the essential oil extraction process have not been evaluated for their potential as a spent substrate for enzyme production, despite other tropical resources such as rice cane (*Oryza sativa*), banana stems (*Musa paradisiaca*), wheat straw (*Triticum* spp.), sugar cane (*Saccharum officinarum*), and olives (*Olea europea*) [27] being utilized. Furthermore, in Ecuador, only two fungi, namely *Xylaria guianensis* (Mont.) Fr. and *Lentinula edodes* (Berk.) Pegler were studied for enzyme production, in which the former didn't show any enzymatic activity [28,29], which underlines the need for further biodegradation experiments with other *Xylaria* spp. to be carried out.

The overall objective of this study was to assess the possible biotechnological use of the *Xylaria* spp. using residues obtained from the Palo Santo essential oil extraction (BGR's). For that, the degradation dynamics of lignin and cellulose of the BGR's were analyzed by two native fungi (*Xylaria* spp.) of the dry forest areas of southern Ecuador, and compared to the behavior of *Trametes versicolor* (L.) Lloyd, due to its well-known degradation and enzyme production capacities. Furthermore, changes in the carbon, nitrogen, phosphorus, and potassium contents of the BGR's during the degradation process were evaluated.

2. Materials and Methods

2.1. Identification of Fungi

The fruiting bodies of *Xylaria* spp. were collected on stumps of dead wood of *B. graveolens* trees in the tropical dry forest of southern Ecuador (Figure 1). The samples were analyzed and deposited in the Herbarium of the Technical University of Loja (HUTPL), Fungarium section, using taxonomic criteria [30,31]. The fungus used as control was *Trametes versicolor* (isolated on MEA: Malt Extract Agar), which was donated by the Spanish National Research Council (Consejo Superior de Investigaciones Científicas, CSIC). This fungus was selected as a model due to its high enzymatic activity, especially in laccase, xylanase, and cellulose, during the degradation process [27].

2.2. Isolation, DNA Extraction, PCR and Fungal Sequencing

After the taxonomic identification of the *Xylaria* spp. the samples were disinfected with 5% sodium hypochlorite solution for three minutes. To flush out the solution and to clean the samples, distilled water was used for three minutes. Then, a second disinfection was applied with a 70% ethanol solution for one minute. Finally, the samples were cleaned with distilled water for one minute again [20].

The disinfected ascomes of each fungus were placed separately into Petri dishes on MEA (malt extract Agar), where the cell cultures were incubated at 25 °C until the mycelia growth was completed (after seven days). After this, the individual fungi were extracted and placed into Petri dishes on MEA again to guarantee the purity of the fungal isolation.

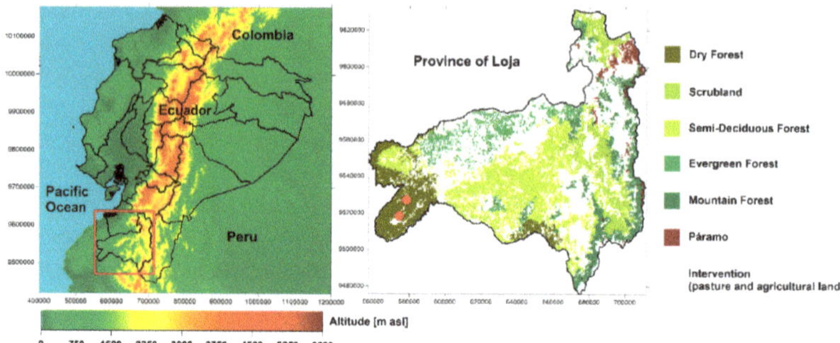

Figure 1. Digital Elevation Model (DEM) of continental Ecuador (left) and natural ecosystems of the province of Loja (right). The map was adapted from the Ecuadorian Ministry of the Environment [32]. The red symbols indicate the sampling point of the *Xylaria* spp.

To be sure that the individual fungal isolation only contained the required fungus species, a DNA extraction (PCR and sequencing test) was executed, following the protocol described by Iotti and Zambonelli [33] and Tamura et al. [34]. The phylogenetic location of the isolates was established by morphological observations and the DNA sequence of the ITS-5.8S region. As universal primers ITS1/LR5 or ITS1/NL4: ITS1 5'TCC GTA GGT GAA CCTGGG 3' [35], LR5 5'TCC TGAGGG AAA CTT CG 3' [36], NL4 5'GGT CCG TGT TTC AAG ACGG 3' [35] were used to amplify the ITS-5.8S region. The DNA sequences of the *Xylaria* spp. were classified by means of the registered species in the GenBank database using BLAST searching (https://www.ncbi.nlm.nih.gov; see also (supplementary material).

2.3. Preparation and Chemical Analysis of the Spent Substrate

The spent substrate used for the degradation experiment were the fruit residues from the essential oil extraction of *B. graveolens* (BGR's). The BGR's mainly consists of the skin and seeds of the fruits, which contain fiber, water, and fatty acids [5]. The residues were obtained from the UTPL Natural Products Institute (Loja, Ecuador), where the majority of the country's essential oil is produced.

The raw BGR's were dried at 60 °C for 48 h to avoid contamination and biodeterioration [37]. Then, the substrate was sieved through a 2-mm mesh and the pH determined. The contents of acid detergent fiber (ADF), acid detergent lignin (ADL) and cellulose in the BGR's were analyzed according to the method of Van Soest [38]. Briefly, the samples were digested with cetyl trimethyl ammonium bromide (CTAB) for 1 h at 150 °C and afterwards the residues washed with distilled water and then filtered (Whatman GF/C) using a vacuum pump. The retained residues in the filter were oven dried and weighed to calculate the ADF and the weight of the residues after the digestion were compared to the original weight and the difference determined. Then, a second digestion of the residues was applied using H_2SO_4 over 3 h at 25 °C. Afterwards, the residues were washed several times with distilled water to remove the excess acid, and then oven dried for one day at 105 °C and weighed again. The ADL was calculated by means of the percentage of the residues after the second digestion compared to the weight of the material after the first digestion. Finally, residues of the second digestion were incinerated in an oven for a period of 5 h at 500 °C to estimate the ash content. The percentage of cellulose was estimated by the difference between ADF and ADL [39].

The total carbon (TC) and total nitrogen (TN) contents of the BGR's were measured using an auto-analyzer CHNS (Elemental Thermo Finnigan Flash EA1112 CHNS-O). The total phosphorus (TP) and total potassium (TK) contents were determined after the acid digestion following the methodology proposed by Sommer and Nelson [40]. Briefly, 200 mg of the crushed BGR's samples were mixed with 5 mL of a perchloric acid solution (60%) and nitric acid (60%) in relation 3:5 (v/v), which was executed in a BD-40 digester block divided in two phases: 90 min at 130 °C and 75 min at 204 °C. Finally, TP and TK were measured by means of acid extractions in the Agilent 750 Series ICP-MS kit.

2.4. Solid State Fermentation

For the solid state fermentation experiment, 20 g of the BGR's was added into 250 mL Erlenmeyer flasks and mixed with 80 mL of distilled water. To evaluate the biodegradation and enzymatic activities over time by means of the fermentation of the BGR's, different test series of the two selected native fungi (*Xylaria* spp.) and the control fungus *T. versicolor* were prepared (20 Erlenmeyer flasks for each fungus type).

Before inoculating the fungi in their specific flask, the mycelium grew for 15 days on MEA, and all flasks containing the BGR's were sterilized in a Gemmy SA-300VF autoclave. After cooling the flasks, the individual fungi were inoculated using 1 cm^2 of each cell culture. Then, the flasks and their contents were incubated at a temperature of 25 °C, following the method proposed for *Xylaria* spp. by Liers et al. [41] and Rodrigues Negrão et al. [22].

The solid-state fermentation experiment lasted 60 days and samples were analyzed on day 7, 15, 30 and 60 after incubation. At each sampling day five recipients of each fungus, randomly chosen, were selected and examined, applying the quartering method [42]. Then, the five selected samples of each fungus were transferred to other recipients for lyophilization. Lyophilization was executed at −4 °C and 0.1 mm Hg using a LABCONCO equipment [43].

2.5. Chemical Analysis and Quantification of BGR's Degradation

After lyophilization, the BGR's were analyzed to quantify the degradation kinetics at each sampling day (7, 15, 30 and 60), applying the same methods related to the preparation and chemical analysis of the spent substrate. In the process, the degradation ratio of lignin and cellulose was calculated using the following equation [44]:

$$Ri\ (\%) = 100 * (m_o - m_i)/m_o \tag{1}$$

where Ri is the percentage of degradation for the sampling of the week; mo is the initial content of lignin and mi is the content of lignin sampling for each week of degradation.

Furthermore, the reduction in total carbon (TC), total nitrogen (TN), total phosphorus (TP), and total potassium (TK) of the BGR's was determined.

2.6. Enzymatic Assay

The lyophilized samples were mixed with distilled-deionized water, filtered and centrifuged for 30 min at 7 °C and 8500 rpm, before the enzymatic analysis [45]. The laccase activity (E.C.1.10.3.2; p-diphenol: dioxygen oxidoreductase) was determined by the oxidation of syringaldazine (4-hidroxi-3.5-dimetoxibenzaldehidacine) in a 0.22 mM methanolic solution [45,46]. The reaction was carried out at pH 6.5 to measure the corresponding quinone at a wave length of 530 nm ($\varepsilon = 65{,}000$ M^{-1} cm^{-1}). The enzymatic activity of laccase was defined by the amount of enzyme that catalyzed the transformation of 1 μmol substrate per minute.

The xylanase activity was determined by measuring the reduction of glucose. Therefore, a 1% xylene solution in an acetate buffer (50 mM) was used [47,48]. The reaction was executed at a pH of 5.0 and measured at a wave length of 575 nm. To quantify the enzymatic activity, the amount of enzyme that reduced 1 μmol of xylose per minute was determined.

The cellulase activity (E.C.3.2.1.4; β-1,4-endoglucanase) was determined by measuring the formation of reduced sugar (glucose) [48,49] using carboxymethylcellulose (1%) in an acetate buffer (50 mM) as substrate. The reaction was executed at a pH of 5.0 and the product measured at a wave length 550 nm. The enzymatic activity was defined by the amount of enzyme that produced 1 μmol of glucose per minute.

In this study, the enzyme activity is expressed in terms of the volumetric activity in Units per Liter (U/L). To obtain the activity value for each enzyme (laccase, xylanase and cellulase) and for each fungus on the respective sampling day, the arithmetic mean of the five analyzed samples was calculated.

2.7. Statistical Analysis

The degradation capacity of the studied fungi inoculated on the BGR's was evaluated by a one-way ANOVA using the SPSS Statistical Software package (v.15.0; SPSS Inc., Chicago, USA). The correlations between the measured variables were determined by the Pearson correlation coefficient; significance was accepted at p-value < 0.05 in all cases. The enzymatic activity of the fungi during the solid-state fermentation experiment was assessed through a repeated measures ANOVA. The differences between the means were evaluated through the multiple range Tukey Test (HSD) and accepted at a significance level of 5% (p-value < 0.05).

3. Results and Discussion

3.1. Identification of Fungi

The sequences of the utilized *Xylaria* sp. were compared to the species registered in the international GenBank (https://www.ncbi.nlm.nih.gov) to realize the phylogenetic analysis (Table 1).

Table 1. Of accessions in the GenBank of species used for phylogenetic analyses (https://www.ncbi.nlm. nih.gov).

Species Isolates	Collection Number and DNA Region	GenBank Accession No.	Country	Reference
Xylaria feejeensis	EGJMP22 ITS-5.8S	KF192827	India	Jagan et al. [50]
Xylaria feejeensis	EGJMP30 ITS-5.8S	KF177680	India	Jagan et al. [50]
Xylaria feejeensis	HMJAU 22039 ITS-5.8S	JX256824	China	Ma et al. [51]
Xylaria feejeensis	A2S4-D46 ITS-5.8S	KJ767110	Malaysia	Teh & Latiffah. [52]
Xylaria feejeensis	A1S3-D88 ITS-5.8S	KJ767104	Malaysia	Teh & Latiffah. [52]
Xylaria feejeensis	Genes; ITS-5.8S	AB569622	Japan	Siriwach et al. [53]
Xylaria feejeensis	Genes; ITS-5.8S	AB809464	Tailandia	Srisapoomi et al. [54]
Xylaria feejeensis	E6912b ITS-5.8S	HM992808	USA	Bascom-Slack et al. [55]
Xylaria feejeensis	1012 ITS-5.8S	GU322460	—	Hsieh et al. [56]
Xylaria feejeensis	860 ITS-5.8S	GU991523	—	Hsieh et al. [56]
Xylaria curta	SGLAf81 ITS-5.8S	EU715634	—	Soca-Chafre et al. [57]
Xylaria hypoxylon	95082001 ITS-5.8S	GU300095	—	Hsieh et al. [56]
Xylaria grammica	152 ITS-5.8S	KF312440	—	Jagan et al. [50]
Xylaria bambusicola	162 ITS-5.8S	GU300088	—	Hsieh et al. [56]
Xylaria bambusicola	205 ITS-5.8S	EF026123	—	Hsieh et al. [56]
Xylaria venosula	94080508 ITS-5.8S	EF026149	—	Hsieh et al. [56]
Xylaria venosula	ITS-5.8S	JN198529	—	Wu et al. [58]
Xylaria microceras	414 ITS-5.8S	GU300086	—	Hsieh et al. [56]
Collodiscula japonica	CJ ITS-5.8S	JF440974	Austria	Jaklitsch & Voglmayr [59]
Xylaria cf. *microceras*	1m_VC ITS-5.8S	KT250967	Ecuador	Carrión-Paladines et al. [60]
Xylaria feejeensis	Cb_VC4 ITS-5.8S	KT250968	Ecuador	Carrión-Paladines et al. [60]
Xylaria feejeensis	Cn_VC3ITS-5.8S	KT250969	Ecuador	Carrión-Paladines et al. [60]
Xylaria cf. *microceras*	VCF1 ITS-5.8S	KT250970	Ecuador	Carrión-Paladines et al. [60]
Xylaria feejeensis	VCF10 ITS-5.8S	KT250971	Ecuador	Carrión-Paladines et al. [60]
Xylaria feejeensis	VCF3c ITS-5.8S	KT250972	Ecuador	Carrión-Paladines et al. [60]
Xylaria feejeensis	VCF4c ITS-5.8S	KT250973	Ecuador	Carrión-Paladines et al. [60]
Xylaria cf. *microceras*	VCF7 ITS-5.8S	KT250974	Ecuador	Carrión-Paladines et al. [60]
Xylaria cf. *microceras*	VCF7c ITS-5.8S	KT250975	Ecuador	Carrión-Paladines et al. [60]
Xylaria feejeensis	VCF9c ITS-5.8S	KT250976	Ecuador	Carrión-Paladines et al. [60]
Xylaria cf. *microceras*	xml_ VC8 ITS-5.8S	KT250977	Ecuador	Carrión-Paladines et al. [60]

The results are shown in Figure 2, where two clades are presented, which indicate an accordance of over 70%, when analyzing the ITS-5.8S regions. By means of the GenBank sequences, one fungus used in this study was identified as *X. feejeensis*, whereas the other native fungus could not be classified precisely. The closest sequence was related to *X. microceras* (accession code GU300086), because the morphology was similar but the genetics different. Therefore, this fungus is probably a new species, which should be analyzed more systematically. Consequently, this fungus is named *X.* cf. *microceras* for the present study.

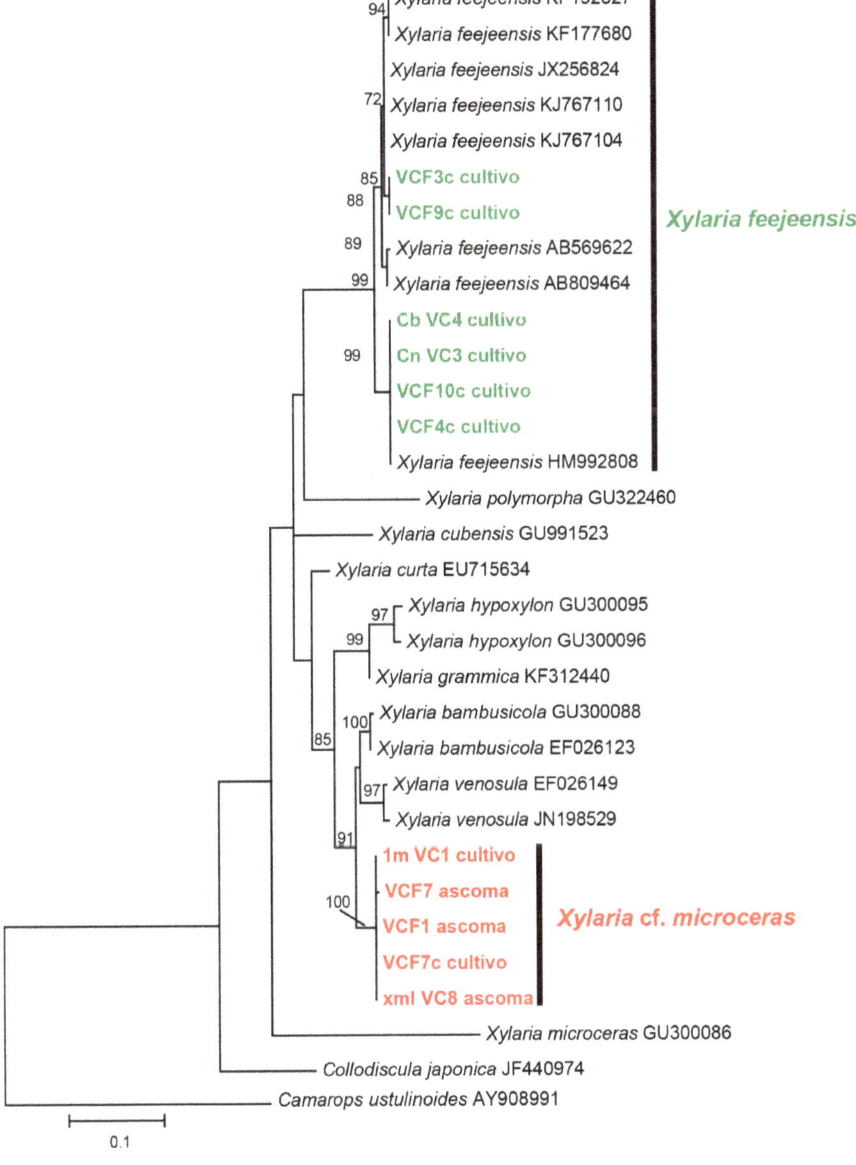

Figure 2. Phylogenetic location of *X. feejeensis* (green) and *X.* cf. *microceras* (red) based on our ITS-5.8S sequences (in bold) and the most related sequences from the GenBank. Only values ≥ 70% are shown on the nodes. The sequence from *Camarops ustulinoides* with accession number AY908991 was used as out group.

3.2. Spent Substrate Characterization

Table 2 shows the chemical characterization of the BGR's before the biodegradation by the three fungus species. The spent substrate had an average lignin and cellulose content of 9.1% and 19.5%, respectively. The mean TC and TN were 47.3% and 1.5 %, which resulted in an average C/N ratio of 30.4. The TP and TK contents were 0.2% and 1.2% respectively. Sulfur (S) was absent in the BGR's, which indicates that the substrate is suitable for biodegradation because no SO_2 (reactive gas) can be emitted [61]. The average pH value of the spent substrate was 7.0.

Table 2. Biochemical properties of BGR's in natural form, prior to inoculation with the strains (*X. feejeensis*, *X.* cf. *microceras* and *T. versicolor*) and their standard deviation based on four replicates.

Biochemical Compound	Average	Standard Deviation
Lignin (%)	9.1	0.79
Cellulose (%)	19.5	0.91
TC (%)	47.3	0.20
TN (%)	1.5	0.08
C/N ratio	30.4	1.57
TP (%)	0.2	0.02
TK (%)	1.2	0.20
S (%)	0.0	0.00
pH	7.0	0.16

The lignin content is important for methane (biogas) production [62], where garden waste (lignin: 10.5%), rice straw (10.8%), shells of *Durio zibethinus* (11.4%) and vinegar residues (lignin: 12.4%) are generally used [63]. The lignin content of the BGR's (9.1%, Table 2) is slightly lower than these substrates, but higher compared to other wastes, which are also used for methane production, such as the leaves and seeds of *Chenopodium album* (lignin: 7.7%), its fruit and vegetable (7.9%), and seeds of *Durio zibethinus* (8.8%) [63,64], which makes the BGR's suitable for methane (biogas) production.

The cellulose content is important for ethanol production [65], where generally banana peel and skin (13.2% and 9.2%, respectively) as well as rice bran are used. The BGR's (19.5%, Table 2) had notably higher cellulose content, which indicates that ethanol can be potentially produced from the BGR's. However, for the manufacture of cosmetics, higher cellulose contents are required [24], which can be found in fiber sorghum (*Sorghum* sp. 41.8%) and in rice husk (*Oryza sativa*, 33.0%) [66].

The TC (47.3%) and TN (1.5%) content establish the C/N ratio, which is important for biodegradation experiments, because substrates with high C/N ratios usually produce the immobilization of nutrients during the process [25]. The optimal range of the C/N ratio lies between 20 and 30, as Montingelli et al. [67] stated. For the BGR's, a nearly optimal C/N ratio (30.4) was obtained, although the TC and TN contents were lower compared to other wastes used for oil extraction, like residues of olive fruits (*Olea europaea*; 58.5% TC, 1.8 % TN, C/N ratio: 31) or the seeds of the litchi (*Litchi chinensis* 56.1% TC, 1.1% TN; C/N ratio: 51) [68]. However, the C/N ratio of the BGR's is much more appropriate than these spent substrates, as well as other spent substrates (e.g., rice straw, *Oryza sativa*; TC: 57.7, TN 0.5%, C/N ratio: 115.0) used in biodegradation experiments [69], which underlines the utility of the BGR's. Furthermore, Motingelli et al. [67] found, that the maximum methane yield is produced when the C/N ratio is around 30, which additionally affirms the potential use of the BGR's for biogas production.

The contents of TP and TK (Table 2) can restrict the degradation of the spent substrate, because these nutrients have an influence on the physiology and the growth of the fungi [26,70,71]. The BGR's showed an average TP content of 0.2%, which is similar to olive residues and rice straw [68]. However, according to El-Haddad et al. [69], the optimal TP values range between 0.7% and 1.1%, which indicates a deficiency of this element in the BGR's. The optimal range of TK lies between 1% and 3% [59], which indicates that the TK content of the BGR's (1.2%, Table 2) is at the lower end of the optimal range, but still adequate for biodegradation experiments. Finally, the neutral pH of the BGR's favors the development of the fungi, because pH values around 7 increase their growth, especially for *Xylaria* spp. [72].

3.3. Efficiency of Degradation of the Three Fungi in the BGR's

Figure 3 shows the degradation of lignin (a) and cellulose (b) of the BGR's in the presence of the fungi. The degradation increased for all fungi during the incubation period, reducing the original content of lignin between 15% and 34% and of cellulose between 28% and 56%.

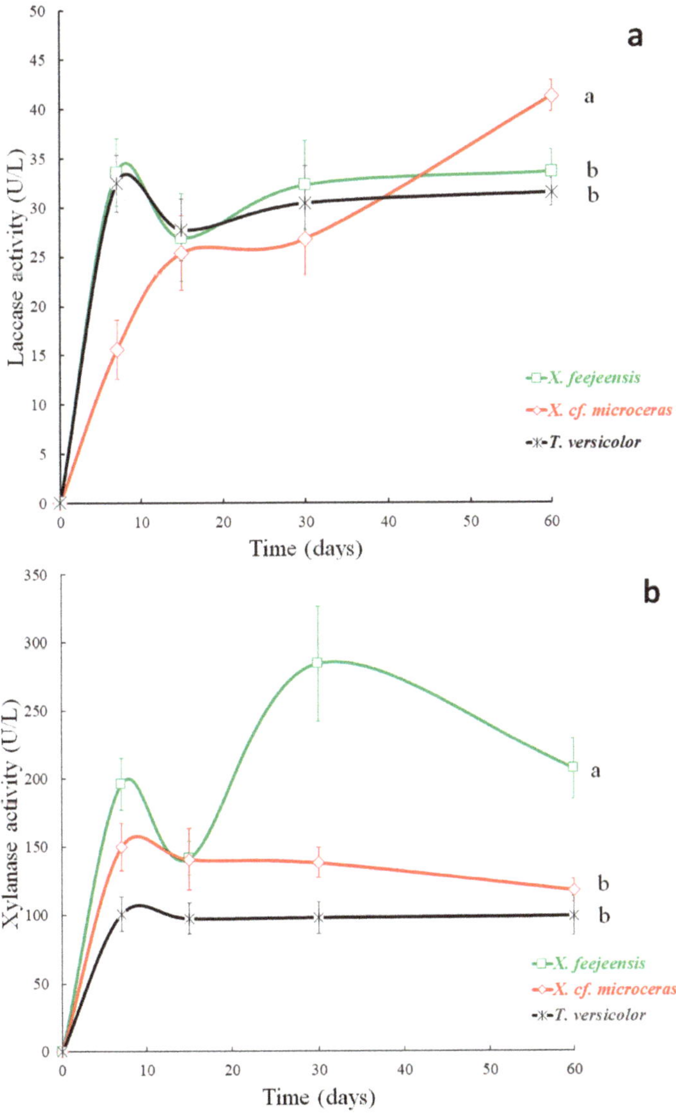

Figure 3. Lignin (**a**) and cellulose (**b**) degradation of BGR's by native fungi of *Xylaria* spp. and *T. versicolor* in solid state fermentation. The bars show the standard error of the mean measured activity of the four samples analyzed on each sampling day. Different letters (**a**,**b**) indicate significant difference ($p \leq 0.05\%$, HSD Tukey).

However, significant differences were found between the fungus types, at which *T. versicolor* was the most efficient fungus for lignin degradation (33.8%, Figure 3a), but the poorest for cellulose degradation (28.3%, Figure 3b). The two native fungi isolated from the Palo Santo wood, *X. cf. microceras* and *X. feejeensis*, were less effective in the degradation of lignin (Figure 3a), but demonstrated their capacity, which is also confirmed by Osono and Takeda [17] and Koide et al. [18]. These studies illustrated that *Xylaria* spp. have the potential to degrade lignin as well as holocellulose, because they produce selective delignification and therefore have a good ligninolytic capacity. Furthermore, the high degradation capacity of *Xylaria* spp. is reported by Pointing et al. [73], Chaparro et al. [20] and Rodrigues Negrão et al. [22], who consider *Xylaria* spp. within the group of white-rot fungi, which degrade lignin effectively.

In contrast, *X. cf. microceras* was the most effective fungus for cellulose degradation, reducing the original cellulose content of the BGR's by about 56.6% during the incubation period (Figure 3b), followed by *X. feejeensis* (42.3%) and *T. versicolor* (28.3). The relatively high degradation rate of the two *Xylaria* spp. are consistent with findings of previous studies [73], where it was stated that all Xylariaceae taxa have high capacities to hydrolyze lignocellulosic resources.

3.4. Mineralization of TC, TN and Evaluation of C/N Ratio

During degradation of the organic matter, organic carbon is needed by the fungi as an energy and biomass source, converting it partially into CO_2 and under certain circumstances also into methane (CH_4). The organic nitrogen is mainly transformed into available N (nitrate and ammonium) by the fungi during the decomposition, and afterwards partially assimilated to build new biomass [74]. Therefore, after 60 days of incubation, the initial TC content of the BGR's was strongly reduced, whereas the original TN content of the BGR's (1.5%, Table 1) showed only small variations during the whole solid state experiment (Table 3).

The TC mineralization was highest during the first seven days of incubation for all three fungi, when almost 50% of the initial TC content was degraded. However, TC content was most effectively degraded by *T. versicolor* (final content: 16.1%) followed by *X. feejeensis* (final content: 17.0%) and *X. cf. microceras* (final content: 19.1%), which can be traced back to the high secretion of cellulolytic enzymes of all fungi during the experiment, because the spent substrate is relatively rich in organic carbon.

The TC mineralization was positively correlated to the degradation of lignin (0.58, $p < 0.01$) and higher to the degradation of cellulose (0.82, $p < 0.01$; Table 4). In addition, the correlation between the reduction of lignin or cellulose and the decrease of the TC content in the BGR's was significant, which was also found in other investigations [75].

TN content remained stable (~1.4%) during the incubation period for the three fungi (Table 3). The small variations in TN content during the mineralization process can be explained by the moderate TN content of the BGR's. The fungi mainly used the carbon content as an energy source and to build biomass. This finding is confirmed by Rigby et al. [76], who stated that the TN content of the spent substrate is only degraded if it is needed for the metabolic requirements of the fungi (microbial cells). In this case, the TC content of the BGR's is sufficient for the metabolic requirement and development of the fungi, while the TN content stayed more or less stable during the whole incubation period. Besides this, the moderate TN content of the BGR's (initially; 1.5%) makes the substrate suitable as organic fertilizer, because, according to the European eco-label, an organic fertilizer should not exceed 3% of TN.

As expected, the C/N ratio decreased notably during the inoculation period (Table 3) because of the degradation of the TC content of the BGR's [77]. The C/N ratio was positively correlated with the degradation of lignin ($r = 0.56$, p-value < 0.01) and cellulose ($r = 0.84$, p-value < 0.01), which underline the relation between TC degradation and lignin/cellulose reduction (Table 4).

Table 3. Changes in biochemical constituents of Palo Santo waste (*B. graveolens*) during 60 days of solid-state fermentation, with native fungi *Xylaria* spp. and *T. versicolor*.

Parameter	*X. feejeensis*				*X. cf. microceras*				*T. versicolor*			
	Days of Incubation				Days of Incubation				Days of Incubation			
	7	15	30	60	7	15	30	60	7	15	30	60
TC (%)	28.4 ± 0.8	24.2 ± 1.0	20.7 ± 1.1	17.0 ± 0.4	28 ± 1.0	25.4 ± 1.2	21.7 ± 1.2	19.1 ± 1.1	26.7 ± 0.4	22.7 ± 0.7	19.6 ± 0.8	16.1 ± 0.7
TN (%)	1.4 ± 0.0	1.4 ± 0.1	1.5 = 0.0	1.4 ± 0.1	1.3 ± 0.1	1.4 ± 0.0	1.4 ± 0.1	1.4 ± 0.1	1.3 ± 0.0	1.4 ± 0.1	1.4 ± 0.1	1.4 ± 0.0
C/N ratio	20.7 ± 0.4	17.2 ± 1.3	14.1 ± 1.0	11.9 ± 0.7	21.4 ± 1.8	18.4 ± 0.6	15.0 ± 0.7	13.3 ± 0.9	21.1 ± 0.4	16.6 ± 1.0	13.8 ± 0.8	11.7 ± 0.5
TP (%)	0.2 ± 0.0	0.2 ± 0.0	0.3 = 0.0	0.3 ± 0.0	0.2 ± 0.0	0.3 ± 0.0	0.3 ± 0.0	0.3 ± 0.0	0.2 ± 0.0	0.3 ± 0.0	0.3 ± 0.0	0.3 ± 0.0
TK (%)	0.8 ± 0.1	0.5 ± 0.1	0.7 = 0.0	0.8 ± 0.1	0.7 ± 0.1	0.5 ± 0.1	0.7 ± 0.0	0.8 ± 0.1	0.7 ± 0.1	0.6 ± 0.1	0.7 ± 0.1	0.8 ± 0.0

Table 4. Pearson correlation coefficient among BGR's properties. Significant correlation is shown at $p < 0.05$ (*) and $p < 0.01$ (**).

	Cellulose (%)	TC (%)	TN (%)	C/N	TP (%)	TK (%)
Lignin loss (%)	0.457 **	0.581 **	−0.377 **	0.564 **	−0.265 *	0.164
Cellulose (%)		0.821 **	−0.683 **	0.835 **	−0.372 **	0.383 **
TC (%)			−0.666**	0.967 **	−0.446 **	0.442 **
TN (%)				−0.827 **	0.550 **	-0.636 **
C/N					−0.516 **	0.571 **
TP (%)						−0.385 **

3.5. Mineralization of Phosphorus and Potassium

The TP content of the three test series slightly increased from 0.2% to 0.3% during the solid state experiment (Table 3). This is caused by the transformation of the organic phosphorus (Po) into its inorganic form (Pi) [78]. The soluble Pi is afterwards incorporated into the OM of the fungi to build up biomass. These results coincide with those reported by Kuehn and Suberkropp [79], who also observed an increase of TP when inoculating different fungi in decaying litter of the *Juncus effusus*. The optimal range of TP content in organic fertilizer lies between 0.15% and 1.5%, which makes the BGR's an adequate resource for soil improvers.

The initial TK content of the BGR's (1.2%, Table 2) was reduced to 50% after the first 15 days of incubation by all three fungi, and afterwards the values increased again, reaching the final value of approximately 0.8% on the last sampling day (Table 3). Potassium is a very mobile and unstable nutrient, which is needed by the fungi particularly at the beginning of the degradation process [80]. However, the increase of TK at the end of the incubation period was probably due to the mineralization of the OM and the production of CO_2 by the fungi. The typical range of potassium in organic fertilizers is 0.4-1.6%, which indicates that the BGR's do not present a deficiency in this nutrient. Furthermore, the normal TK content of the BGR's induces a good C/N balance, because TK plays an important role in carbon (C) and nitrogen metabolism (N) [81].

3.6. Enzyme Activities

The degradation of lignin and cellulose is a consequence of the enzymatic activity of the fungi, which is mainly caused by the production of the enzymes laccase, xylanase and cellulase [44,82]. Figure 4 shows the variation in laccase (a), xylanase (b) and cellulase (c) activities for the individual fungus types during the incubation period.

3.6.1. Laccase Activity

Laccase activity was detected for all three fungi on the first sampling day (Figure 4a), which was probably due to the absence of sulfur (S) and the moderate contents of TC and TN of the BGR's (Table 2), facilitating the immediate decomposition of lignin [83]. The temporal variability of laccase activity of *X. feejeensis* and *T. versicolor* were similar, increasing notably during the first seven days of incubation (33.5 U/L and 32.4 U/L; respectively), and afterwards remaining more or less stable until the end of the incubation period (final values: of 33.7 U/L, *X. feejeensis*; 31.5 U/L, *T. versicolor*). *X.* cf. *microseras* showed a different behavior, especially during the first seven days of incubation, when the lowest laccase activity of all three fungi was detected (15.6 U/L), and during the end of the incubation period, when the laccase activity of *X.* cf. *microceras* increased notably, reaching 41.3 U/L, which was the highest value of all test series during the complete observation period.

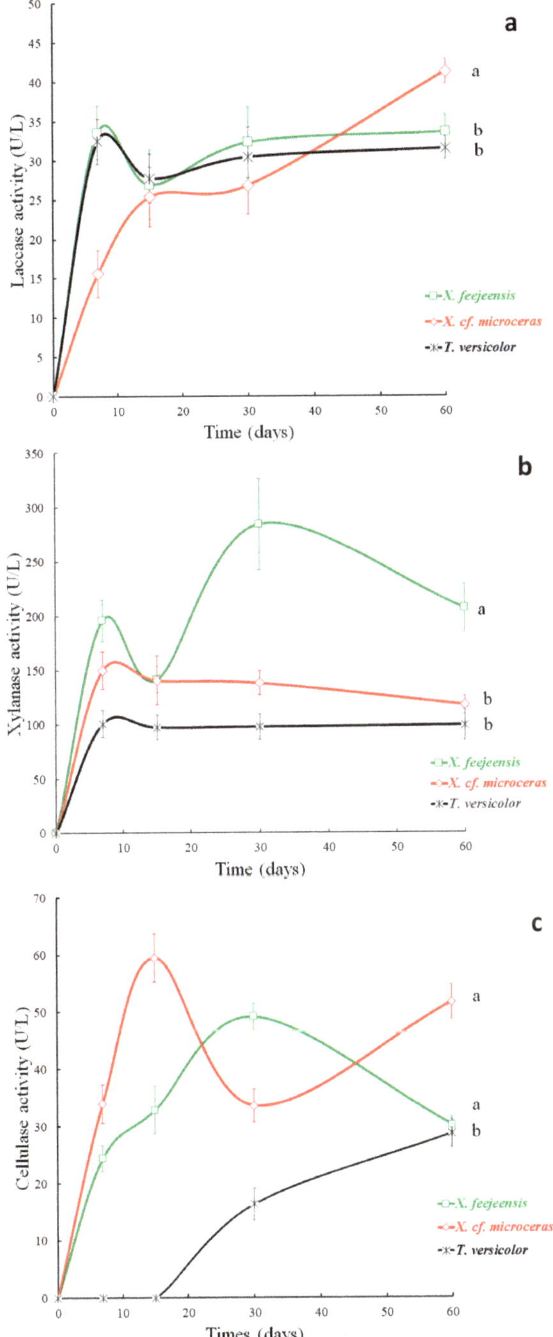

Figure 4. Laccase (**a**), xylanase (**b**) and cellulase (**c**) activities in BGR's with native species of *Xylaria* spp. and *T. versicolor* incubated at 25 °C. The bars show the standard error of the mean measured activity of the four samples analyzed on each sampling day. Different letters indicate significant difference ($p \leq 0.05\%$, HSD Tukey).

However, the variability in the production of the laccase enzyme was expected, because, as Dong et al. [44] showed, laccase acts synergistically with other lignin-degrading enzymes such as polyphenol oxidase (PPO) and manganese peroxidase (MnP).

The observed laccase activity of the three studied fungi seems to be low in comparison to other studies [41], but these investigations applied liquid fermentation (liquid spent substrates) and used 2.5-xylidine or veratryl alcohol as an enzymatic inductor, which increases the laccase activity notably [84]. The solid-state experiment of this study did not utilize any enzyme inductor, which explains the relatively low laccase activity of *T. versicolor* and the *Xylaria* spp. However, investigations using solid substrates obtained similar activity values as observed here [20,85]. Generally, all *Xylaria* spp. have an unusually high ability to degrade lignin compared to other Ascomycota, as Liers et al. [23] and Rodrígues Negrão et al. [22] stated, because *Xylaria* spp. mineralize lignin almost as efficiently as the aggressive fungi of white-rot (e.g., *T. versicolor*).

Besides the solid fermentation applied, the moderate laccase enzyme activity might be due to the low lignin content of the BGR's (9.1%; Table 2), which cause a reduction in laccase production. As Coronel et al. [85] indicated, the enzymatic activity is highly influenced by the chemical composition of the substrate where the fungus was incubated. Therefore, average laccase activity of all fungi was more or less similar (*X. feejeensis*: 30.0 U/L; *T. versicolor*: 29.5 U/L; *X.* cf. *microceras*: 26.6 U/L). However, the statistical analysis showed that a significant difference between *X.* cf. *microceras* and the other two fungi (*X. feejeensis* and *T. versicolor*) respective to laccase activity exists (p-value < 0.05; Figure 4a, different letters), whereas no difference was found between *X. feejeensis* and *T. versicolor* (Figure 4a, similar letters).

3.6.2. Xylanase Activity

Xylanase also act synergistically with other hydrolytic enzymes, such as esterase, to modify the structural configuration of lignocelluloses [44], which is necessary to make the cellulose accessible for degradation. For example, esterase, in combination with xylanase, cleaves covalent bonds between polysaccharides or hemicelluloses, and therefore plays a key role in the degradation of the hemicellulose matrix [82]. The preliminary degradation of the lignin-hemicellulose matrix is proven by the present study because xylanase activity was detected in all three fungi species on the first sampling day, reaching values between 100.7 U/L and 196.0 U/L (Figure 4b).

During the complete observation period, the xylanase activity of *X. feejeensis* was always highest compared to the other two fungi, except on day 15 when the activity of *X. feejeensis* decreased and similar values for both *Xylaria* spp. were measured (140.8 U/L and 141.6 U/L). Highest activity of *X. feejeensis* was observed on day 30 with 284.3 U/L, but afterwards activity decreased. The xylanase activity of *X.* cf. *microceras* and *T. versicolor* started with lower values (150.3 U/L and 100.7 U/L respectively), but their activity curves showed similar behaviors (Figure 4b), decreasing slightly between day 7 and day 15 after incubation and then remained almost stable during the rest of the observation period, reaching final values of 117.9 U/L (*X.* cf. *microceras*) and 99.0 U/L (*T. versicolor*). Statistically significant differences were found between *X. feejeensis* and the other two fungi species, but differences between *X.* cf. *microceras* and *T. versicolor* were not significant (Figure 4b).

In general, *X. feejeensis* and *X.* cf. *microceras* showed higher xylanase activity compared to *T. versicolor*, but also to other *Xylaria* spp. which were inoculated on solid-state experiments [23]. All *Xylaria* spp. can degrade hemicellulose effectively [18,19], but the two *Xylaria* spp. studied here showed even higher activity values than those reported in the literature and therefore might be suitable for commercial xylanase production required for industrial processes in the paper, food and wine industry. *T. versicolor* also showed high xylanase activity, although the xylanase production of this fungus species is typically low and an inducible mechanism (enzyme inductor) is needed to increase the activity. Irbe et al. [86] used glycerol alcohol as an inductor and observed a notable increase in xylanase activity and enzyme production. However, in this study no inductor was applied, but xylanase activity was high for all three fungi, which is probably due to the xylan content of the BGR's. Xylan is the second most abundant hemicellulosic polysaccharide in nature and present in the cell walls of the plants [43]. The residues from the Palo Santo essential oil extraction consist mainly in parts of the fresh fruits [5] and therefore the xylan content was not reduced by other degradation processes, which might explain the high xylanase activity of all three fungi.

3.6.3. Cellulase Activity

As expected, cellulase activity was delayed (Figure 4c), because the lignin-hemicellulose matrix had to be degraded first to make the cellulose accessible [44], which indicates that cellulase is an inducible enzyme [87]. As Arantes and Sadler [88] stated, the cellulose regions are tightly packed with lignin and hemicellulose, which is the major contributing factor to cellulose resistance to degradation. Therefore, amorphogenesis, a process consisting of the degradation of lignin and hemicellulose, is needed to liberate the cellulose, which afterwards can be degraded by cellulase.

Cellulase activity of the two *Xylaria* spp. was detected on the first sampling day, in contrast to *T. versicolor*, where significant activity was not measured until day 30 (Figure 4c). The *X. feejeensis* fungus reached its maximum activity (49.2 U/L) on day 30, whereas *X.* cf. *microceras* peaked on day 15 (59.5 U/L), which was concurrently the highest value of all three fungi species during the entire observation period. *T. versicolor* did not show any cellulase activity until day 15 of incubation and afterwards displayed an almost linear increase to its maximum and final value of 28.7 U/L on the last sampling day.

In general, the cellulase enzyme system consists of three types of enzymes; endo-1,4-β-glucanase (cellulase), cellobiohydrolase or exoglucanases (avicelase), and β-glucosidase (cellobiase), which act in conjunction to degrade the cellulose content of the spent substrate [44]. As Du et al. [89] showed, *T. versicolor* first degrades cellulose through the enzyme cellobiase (β-glucosidase), and later in combination which the enzyme cellulase, for which reason a delay in cellulase activity of *T. versicolor* was observed. In contrast, the *Xylaria* spp. produced the enzyme cellulase immediately after incubation to degrade the cellulose content.

As is also shown in Figure 4c, maximum values of cellulase activity of the individual fungi were not reached simultaneously, due to the different metabolic requirements of each fungus. The difference may be due to the fungus type used (Basidiomycota and Ascomycota), because *T. versicolor* is a fungus of white-rot, which is specifically indicated in the degradation of lignin. Furthermore, the lower cellulase activity of *T. versicolor* might be a consequence of the substrate used (solid), because *T. versicolor* can produce up to 100 U/L of cellulase when inoculated in liquid mediums [90].

The differences in the cellulase activity are also depicted by the repeated measures ANOVA, in which no significant differences between *X. feejeensis* and *X.* cf. *microceras* were found, but differences between the *Xylaria* spp. and *T. versicolor* were significant (p-value < 0.05). In general, average cellulase production of the *Xylaria* spp. was notably higher (27.3 U/L to 35.8 U/L) compared to *T. versicolor* (9.0 U/L), especially during the first 30 days of incubation.

The high cellulase activity of the two *Xylaria* spp. is consistent with results from previous investigations comparing the enzymatic production of different fungi species [19,73]. These studies demonstrated that *Xylaria* spp. are potential producers of cellulolytic enzymes, due to their high cellulolytic activity, and therefore can be used in biotechnological applications as well as for industrial purposes. This is confirmed by Gutiérrez-Soto et al. [16], who measured the cellulase activity of *Xylaria* spp. up to 199 U/L. However, these studies incubated the fungi on liquid mediums; in solid spent substrates, the activity is generally lower [23]. The variation in cellulase enzyme activity of *Xylaria* spp. apparently depends on two factors: first, the species [73] and second, the type of substrate [19].

4. Conclusions

The content of lignin and cellulose in the BGR's makes the substrate suitable for biotechnological applications, especially for the production of methane and ethanol. Furthermore, the contents of the macro nutrients were within the optimal range, and therefore the BGR's can be applied to soils as fertilizers.

The BGR's were also suitable for the production of enzymes for industrial purposes by means of fungal degradation. The native *Xylaria* spp. showed generally higher enzymatic activity than the control fungus and were especially practical for the production of xylanase and cellulase.

Supplementary Materials: Supplementary materials can be found at http://www.mdpi.com/2311-5637/5/3/76/s1.

Author Contributions: Formal analysis, V.C.-P. and A.F.; Investigation, V.C.-P.; Methodology, V.C.-P., R.E.C. and R.G.-R.; Supervision, R.G.-R.; Validation, P.P.D.; Writing—original draft, V.C.-P., R.E.C. and R.G.-R.; Writing—review & editing, V.C.-P., A.F., P.P.D. and R.G.-R.

Funding: This research was funded by the Secretary of Science and Technology of Ecuador (SENESCYT—CEREPS—2007) and the Universidad Técnica Particular de Loja (UTPL).

Acknowledgments: We would like to thank the Secretary of Science and Technology of Ecuador (SENESCYT), the Ecuadorian Ministry of the Environment (MAE, research permits 015-IC-UPN-DRLEOZCH-MA and 027-2013-DPL-MA) and the Department of Biological Sciences of the Universidad Técnica Particular de Loja for their support. Special gratitude to the Spanish National Research Council (Consejo Superior de Investigaciones Científicas, CSIC) for their generous contribution. Finally, we would like to thank Gregory Gedeon for text revision.

Conflicts of Interest: The authors declare no conflict of interest.

References

1. Weeks, A.; Tye, A. Phylogeography of palo santo trees (*Bursera graveolens* and *Bursera malacophylla*; Burseraceae) in the Galápagos archipelago. *Bot. J. Linn. Soc.* **2009**, *161*, 396–410. [CrossRef]

2. Carrión-Paladines, V.; García-Ruiz, R. Floristic composition and structure of a deciduous dry forest from southern Ecuador: Diversity and aboveground carbon accumulation. *Int. J. Curr. Res. Acad. Rev.* **2016**, *4*, 154–169. [CrossRef]

3. Yukawa, C.; Imayoshi, Y.; Iwabuchi, H.; Komemushi, S.; Sawabe, A. Chemical composition of three extracts of *Bursera graveolens*. *Flavour Frag. J.* **2006**, *21*, 234–238. [CrossRef]

4. Alonso-Castro, A.; Villareal, M.; Salazar-Olivo, L.; Gomez-Sanchez, M.; Dominguez, F.; Garcia-Carranca, A. Mexican medicinal plants used for cancer treatment: Pharmacological, phytochemical and ethnobotanical studies. *J. Ethnopharmacol.* **2011**, *133*, 945–972. [CrossRef] [PubMed]

5. Carrión-Paladines, V.; Fries, A.; Gómez-Muñoz, B.; García-Ruiz, R. Agrochemical characterization of vermicomposts produced from residues of Palo Santo (*Bursera graveolens*) essential oil extraction. *Waste Manag.* **2016**, *58*, 135–143.

6. Schmid, A.; Hollmann, F.; Park, J.B.; Bühler, B. The use of enzymes in the chemical industry in Europe. *Curr. Opin. Biotechnol.* **2002**, *13*, 359–366. [CrossRef]

7. Johannes, T.; Zhao, H. Directed evolution of enzymes and biosynthetic pathways. *Curr. Opin. Microbiol.* **2006**, *9*, 261–267. [CrossRef] [PubMed]

8. Singh, R.; Kumar, M.; Mittal, A.; Mehta, P.K. Microbial enzymes: Industrial progress in 21st century. *3 Biotech.* **2016**, *6*, 3–15. [CrossRef]

9. Demain, A.; Adrio, J. Contributions of microorganisms to industrial biology. *Mol. Biotechnol.* **2008**, *38*, 41–55. [CrossRef]

10. Trivedi, N.; Reddy, C.R.K.; Radulovich, R.; Jha, B. Solid state fermentation (SSF)-derived cellulase for saccharification of the green seaweed Ulva for bioethanol production. *Algal Res.* **2015**, *9*, 48–54. [CrossRef]

11. Araújo, R.; Casal, M.; Cavaco-Paulo, A. Application of enzymes for textile fibres processing. *Biocatal. Biotransform.* **2008**, *26*, 332–349.

12. Rodríguez Couto, S.; Toca Herrera, J.L. Industrial and biotechnological applications of laccases: A review. *Biotechnol. Adv.* **2006**, *24*, 500–513. [CrossRef] [PubMed]

13. Mainardi, P.; Feitosa, V.; Brenelli de Paiva, L. Laccase production in biorreactor scale under saline condition by the marine-derived basidiomycete *Peniophora* sp. CBMAI 1063. *Fungal Biol.* **2018**, *122*, 302–309. [CrossRef] [PubMed]

14. Gutiérrez, A.; del Río, J.; Martínez, A. Microbial and enzymatic control of pitch in the pulp and paper industry. *Appl. Microbiol. Biotechnol.* **2009**, *82*, 1005–1018. [CrossRef] [PubMed]

15. Trierveiler-Pereira, L.; Romero, A.; Baltazar, J.; Loguercio-Leite, C. Addition to the knowledge of *Xylaria* (Xylariaceae, Ascomycota) in Santa Catarina, Southern Brazil. *Mycotaxon* **2008**, *107*, 139–156. [CrossRef]

16. Gutiérrez-Soto, G.; Medina-González, G.; Treviño-Ramirez, J.; Hernández-Luna, C. Native macrofungi that produce lignin-modifyieng enzymes, cellulases, and xylanases with potential biotechnological applications. *Bioresources* **2015**, *10*, 6676–6689.

17. Osono, T.; Takeda, H. Comparison of litter decomposing ability among diverse fungi in a cool temperate deciduous forest in Japan. *Mycologia* **2002**, *94*, 421–427. [CrossRef]

18. Koide, K.; Osono, T.; Takeda, H. Fungal succession and decomposition of *Camellia japonica* leaf litter. *Ecol. Res.* **2005**, *20*, 599–609. [CrossRef]

19. Bezerra, J.; Santos, M.; Svedese, V.; Lima, D.; Fernandes, M.; Paiva, L.; Souza-Motta, C. Richness of endophytic fungi isolated from *Opuntia ficus-indica* Mill. (Cactaceae) and preliminary screening for enzyme production. *World J. Microbiol. Biotechnol.* **2012**, *28*, 1989–1995. [CrossRef]

20. Chaparro, D.; Rosas, D.; Varela, A. Aislamiento y evaluación de la actividad enzimática de hongos descomponedores de madera (Quindío, Colombia). *Rev. Iberoam. Micol.* **2009**, *26*, 238–243. [CrossRef]

21. Moissenko, K.; Vasina, D.; Farukshina, K.; Savinova, O.; Glazunova, O.; Fedorova, T.; Tyashelova, T. Orchestration of the expression of the laccase multigene family in white-rot basidiomycete *Trametes hirsuta* 972: Evidences of transcription level subfunctionalization. *Fungal Biol.* **2018**, *122*, 353–362. [CrossRef] [PubMed]

22. Rodriguez Negrão, D.; Fernandes da Silva J#xFA;nior, T.; de Souza Passos, J.; Angeli Sansílogo, C.; de Almeida Minhoni, M.; Furtado, E. Biodegradation of *Eucalyptus urograndis* wood by fungi. *Int. Biodeterior. Biodegrad.* **2014**, *89*, 95–102.

23. Liers, C.; Ullrich, R.; Steffen, K.T.; Hatakka, A.; Hofrichllter, M. Mineralization of 14C-labelled synthetic lignin and extracellular enzyme activities of the wood-colonizing ascomycetes *Xylaria hypoxylon* and *Xylaria polymorpha*. *Appl. Microbiol. Biotechnol.* **2006**, *69*, 573–579. [CrossRef] [PubMed]

24. Coseri, S. Cellulose: To depolymerize or not to? *Biotechnol. Adv.* **2017**, *35*, 251–266. [CrossRef] [PubMed]

25. Tognetti, C.; Mazzarino, M.; Laos, F. Improving the quality of municipal organic waste compost. *Bioresour. Thechnol.* **2006**, *98*, 1067–1076. [CrossRef] [PubMed]

26. Vyas, T.K.; Dave, B.P. Effect of addition of nitrogen, phosphorus and potassium fertilizers on biodegradation of crude oil by marine bacteria. *Indian J. Mar. Sci.* **2010**, *39*, 143–150.

27. Ntougias, S.; Baldrian, P.; Ehaliotis, C.; Nerud, F.; Merhautováa, V.; Zervakis, G. Olive mill wastewater biodegradation potential of white-rot fungi—Mode of action of fungal culture extracts and effects of ligninolytic enzymes. *Bioresour. Technol.* **2015**, *189*, 121–130. [CrossRef] [PubMed]

28. Ullah, M.; Camacho, R.; Evans, C.; Hedger, J. Production of ligninolytic enzymes by species assemblages of tropical higher fungi from Ecuador. In *Tropical Mycology*; Watling, R., Frankland, J.C., Ainsworth, A.M., Isac, S., Robinson, C.H., Eds.; CABI Publishing: Wallingford, UK, 2002; Volume 1, pp. 101–112.

29. Vaca, M.; Izurieta, B.; Espín, N. Obtención de extractos enzimáticos con actividad celulolítica y ligninolítica a partir del hongo *Pleurotus ostreatus* 404 y 2171 en rastrojo de maíz. *Revista Politec.* **2014**, *33*, 2.

30. Hladki, A.; Romero, A. A preliminary account of *Xylaria* in the Tucuman Province, Argentina, with a key to the known species from the northern provinces. *Fungal Divers.* **2010**, *42*, 79–96. [CrossRef]

31. Medel, R.; Castillo, R.; Guzmán, G. Adiciones al conocimiento de *Xylaria* (Ascomycota, Xylariales) en México. *Rev. Mex. Micol.* **2010**, *31*, 9–18.

32. Ministerio del Ambiente de Ecuador (MAE). Mapa de Cobertura y Uso de la Tierra. Available online: http://www.ambiente.gob.ec (accessed on 24 April 2017).

33. Iotti, M.; Zambonelli, A. A quick and precise technique for identifying ectomycorrhizas by PCR. *Mycol. Res.* **2006**, *110*, 60–65. [CrossRef] [PubMed]

34. Tamura, K.; Peterson, D.; Peterson, N.; Stecher, G.; Nei, M.; Kumar, S. MEGA5: Molecular evolutionary genetics analysis using maximum likelihood, evolutionary distance, and maximum parsimony methods. *Mol. Biol. Evol.* **2011**, *28*, 2731–2739. [CrossRef] [PubMed]

35. White, T.J.; Bruns, T.; Lee, S.; Taylor, J. Amplification and direct sequencing of fungal ribosomal RNA genes for phylogenetics. In *PCR Protocols: A Guide to Methods and Applications*; Innis, N., Gelfand, D., Sninsky, J., White, J., Eds.; Academic Press: New York, NY, USA, 1990; pp. 315–322.

36. Vilgalys, R.; Hester, M. Rapid genetic identification and mapping of enzymatically amplified ribosomal DNA from several *Cryptococcus* species. *J. Bacteriol.* **1990**, *172*, 4238–4246. [CrossRef] [PubMed]

37. Ruqayyah, T.; Jamal, P.; Alam, Z.; Mirghani, E. Biodegradation potential and ligninolytic enzyme of two locally isolated *Panus tigrinus* strain on selected agro-industrial wastes. *J. Environ. Manag.* **2013**, *118*, 115–121. [CrossRef] [PubMed]

38. Van Soest, P.J. Use of detergents in the analysis of fibrous feeds II: A rapid method for the determination of fibre and lignin. *J. Assoc. Off. Agric. Chem.* **1963**, *46*, 829–835.

39. Goering, H.K.; Van Soest, P.J. Forage fiber analyses (apparatus, reagents, procedures, and some applications). In *Agriculture Handbook No. 379*; U.S. Agricultural Research Service: Washington, DC, USA, 1970.

40. Sommer, L.; Nelson, D. Determination of total phosphorus in soils: A rapid perchloric acid digestion procedure. *Soil Sci. Soc. Am. Proc.* **1972**, *53*, 32–37. [CrossRef]

41. Liers, C.; Ullrich, R.; Pecyna, M.; Schlosser, D.; Hofrichter, M. Production, purification and partial enzymatic and molecular characterization of a laccase from the wood-rotting ascomycete *Xylaria polymorpha*. *Enzyme Microb. Tech.* **2007**, *41*, 785–793. [CrossRef]

42. Roé-Sosa, A.; Estrada, M.; Calderas., F.; Sánchez-Arévalo, F.; Manero, O.; Orta, L.; de Velasquez, M.T. Degradation and biodegradation of polyethylene whit pro-oxidant additives under compost conditions establishing relationships between physicochemical and rheological parameters. *J. Appl. Polym. Sci.* **2015**, *42721*, 1–11.

43. Wei, D.; Chou, H.; Cheng, M.; Chang, S. Purification and characterization of xylanase from *Xylaria regalis*. *Fung. Sci.* **2005**, *20*, 53–59.

44. Dong, X.Q.; Yang, J.S.H.; Zhu, N.; Wang, E.T.; Yuan, H.L. Sugarcane bagasse degradation and characterization of three white-rot fungi. *Bioresour. Technol.* **2013**, *131*, 443–451. [CrossRef]

45. Mata, G.; Savoie, J. Extracelullar enzyme activities in six *Lentinula edodes* strains during cultivation in wheat straw. *World J. Microb. Biot.* **1998**, *14*, 513–519. [CrossRef]

46. Philippoussis, A.; Diamantopoulou, P.; Papadopoulou, K.; Lakhtar, H.; Roussos, S.; Parissopoulos, G.; Papanikolaou, S. Biomass, laccase and endoglucanase production by *Lentinula edodes* during solid state fermentation of reed grass, bean stalks and wheat straw residues. *World J. Microb. Biot.* **2011**, *27*, 285–297. [CrossRef]

47. Sadaf, A.; Khare, S.K. Production of *Sporotrichum thermophile* xylanase by solid state fermentation utilizing deoiled *Jatropha curcas* seed cake and its application in xylooligosachharide synthesis. *Bioresour. Technol.* **2014**, *153*, 126–130. [CrossRef]

48. Miller, G.L. Use of dinitrosalicylic acid reagent for determination of reducing sugar. *Anal. Chem.* **1959**, *31*, 426–428. [CrossRef]

49. Singhania, R.; Saini, J.K.; Saini, R.; Adsul, M.; Mathur, A.; Gupta, R.; Tuli, D.K. Bioethanol production from wheat straw via enzymatic route employing *Penicillium janthinellum* cellulases. *Bioresour. Technol.* **2014**, *169*, 490–495. [CrossRef]

50. The National Center for Biotechnology Information Advances Science and Health by Providing Access to Biomedical and Genomic Information. Available online: https://www.ncbi.nlm.nih.gov/nuccore/KF192827 (accessed on 24 June 2016).

51. The National Center for Biotechnology Information Advances Science and Health by Providing Access to Biomedical and Genomic Information. Available online: https://www.ncbi.nlm.nih.gov/nuccore/JX256824 (accessed on 14 June 2016).

52. The National Center for Biotechnology Information Advances Science and Health by Providing access to Biomedical and Genomic Information. Available online: https://www.ncbi.nlm.nih.gov/nuccore/KJ767110andKJ767104 (accessed on 14 June 2016).

53. The National Center for Biotechnology Information Advances Science and Health by Providing Access to Biomedical and Genomic Information. Available online: https://www.ncbi.nlm.nih.gov/nuccore/AB569622 (accessed on 14 June 2016).

54. The National Center for Biotechnology Information Advances Science and Health by Providing Access to Biomedical and Genomic Information. Available online: https://www.ncbi.nlm.nih.gov/nuccore/AB809464 (accessed on 14 June 2016).

55. The National Center for Biotechnology Information Advances Science and Health by Providing Access to Biomedical and Genomic Information. Available online: https://www.ncbi.nlm.nih.gov/nuccore/HM992808 (accessed on 14 June 2016).

56. The National Center for Biotechnology Information Advances Science and Health by Providing Access to Biomedical and Genomic Information. Available online: https://www.ncbi.nlm.nih.gov/nuccore/GU322460, GU991523,GU300095,GU300088,EF026123,EF026149andGU300086 (accessed on 14 June 2016).

57. The National Center for Biotechnology Information Advances Science and Health by Providing Access to Biomedical and Genomic Information. Available online: https://www.ncbi.nlm.nih.gov/nuccore/EU715634 (accessed on 14 June 2016).

58. The National Center for Biotechnology Information Advances Science and Health by Providing Access to Biomedical and Genomic Information. Available online: https://www.ncbi.nlm.nih.gov/nuccore/JN198529 (accessed on 14 June 2016).

59. The National Center for Biotechnology Information Advances Science and Health by Providing Access to Biomedical and Genomic Information. Available online: https://www.ncbi.nlm.nih.gov/nuccore/JF440974 (accessed on 14 June 2016).

60. The National Center for Biotechnology Information Advances Science and Health by Providing Access to Biomedical and Genomic Information. Available online: https://www.ncbi.nlm.nih.gov/nuccore/KT250967,KT250968,KT250969,KT250970,KT250971,KT250972,KT250973,KT250974,KT250975,KT250976andKT250977 (accessed on 14 June 2016).

61. Zhang, H.; Schuchardt, F.; Li, G.; Yang, J.; Yang, Q. Emission of volatile sulfur compounds during composting of municipal solid waste (MSW). *Waste Manag.* **2013**, *33*, 957–963. [CrossRef] [PubMed]

62. Zhao, J.; Zheng, Y.; Li, Y. Fungal pretreatment of yard trimmings for enhancement of methane yield from solid-state anaerobic digestion. *Bioresour. Technol.* **2014**, *156*, 176–181. [CrossRef] [PubMed]

63. Li, Y.; Zhang, R.; Liu, G.; Chen, C.; He, Y.; Liu, X. Comparison of methane production potential, biodegradability, and kinetics of different organic substrates. *Bioresour. Technol.* **2013**, *149*, 565–569. [CrossRef] [PubMed]

64. Zhao, C.; Yan, H.; Liu, Y.; Huang, Y.; Zhang, R.; Chen, C.; Liu, G. Bio-energy conversion performance, biodegradability, and kinetic analysis of different fruit residues during discontinuous anaerobic digestion. *Waste Manag.* **2016**, *52*, 295–301. [CrossRef]

65. Santos Michel, R.J., Jr.; Canabarro, N.I.; Alesio, C.; Maleski, T.; Laber, T.; Sfalcin, P.; Foletto, E.; Mayer, F.D.; Kuhn, R.C.; Mazutti, M. Enzymatic saccharification and fermentation of rice processing residue for ethanol production at constant temperature. *Biosyst. Eng.* **2016**, *142*, 110–116. [CrossRef]

66. Johar, N.; Ahmad, I.; Dufresne, A. Extraction, preparation and characterization of cellulose fibres and nanocrystals from rice husk. *Ind. Crop. Prod.* **2012**, *37*, 93–99. [CrossRef]

67. Montingelli, M.E.; Tedesco, S.; Olabi, A.G. Biogas production from algal biomass: A review. *Renew. Sustain Energy Rev.* **2015**, *43*, 961–972. [CrossRef]

68. Sampedro, I.; Marinari, S.; D'Annibale, A.; Grego, S.; Ocampo, J.; García-Romera, I. Organic matter evolution and partial detoxification in two-phase olive mill waste colonized by white-rot fungi. *Int. Biodeterior. Biodegrad.* **2007**, *60*, 116–125. [CrossRef]

69. El-Haddad, M.; Zayed, M.; El-Sayed, G.; Hassanein, M.; El-Satar, A. Evaluation of compost, vermicompost and their teas produced from rice straw as affected by addition of different supplements. *Ann. Agric. Sci.* **2014**, *59*, 243–251. [CrossRef]

70. Bullerman, L. Effects of potassium sorbate on growth and aflatoxin production by *Aspergillus parasiticus* and *Aspergillus flavus*. *J. Food Protect.* **1983**, *46*, 940–942. [CrossRef] [PubMed]

71. Eze, J.M. Translocation of phosphate in mould mycelia. *New Phytol.* **1975**, *75*, 579–581. [CrossRef]

72. Hu, Q.; Zhang, G.; Zhang, R.; Hu, D.; Wang, H.; Ng, T. A novel aspartic protease with HIV-1 reverse transcriptase inhibitory activity from fresh fruiting bodies of the wild mushroom *Xylaria hypoxylon*. *J. Biomed. Biotechnol.* **2012**. [CrossRef] [PubMed]

73. Pointing, S.; Parungao, M.; Hyde, K. Production of wood-decay enzymes, mass loss and lignin solubilization in wood by tropical Xylariaceae. *Mycol. Res.* **2003**, *107*, 231–235. [CrossRef] [PubMed]

74. Digby, A.; Gleason, F.; Mcgee, P. Some fungi in the Chytridiomycota can assimilate both inorganic and organic sources of nitrogen. *Fungal Ecol.* **2010**, *3*, 261–266. [CrossRef]
75. Fogarty, W.; Kelly, C. *Microbial Enzymes and Biotechnology*; Fogarty, W.M., Kelly, C., Eds.; Springer: Dordrecht, The Netherlands, 1990; pp. 71–472.
76. Rigby, H.; Clarke, B.; Pritchard, D.; Meehan, B.; Beshah, F.; Smith, S.; Porter, N. A critical review of nitrogen mineralization in biosolids-amended soil, the associated fertilizer value for crop production and potential for emissions to the environment. *Sci. Total Environ.* **2016**, *541*, 1310–1338. [CrossRef]
77. Fioretto, A.; Di Nardo, C.; Papa, S.; Fuggi, A. Lignin and cellulose degradation and nitrogen dynamics during decomposition of three leaf litter species in a mediterranean ecosystem. *Soil Biol. Biochem.* **2005**, *37*, 1083–1091. [CrossRef]
78. Ngo, P.; Rumpel, C.; Ngo, Q.; Alexis, M.; Velásquez Vargas, G.; Mora Gil, M.; Dang, D.; Jouquet, P. Biological and chemical reactivity and phosphorus forms of buffalo manure compost, vermicompost and their misture with biochar. *Bioresour. Technol.* **2013**, *148*, 401–407. [CrossRef] [PubMed]
79. Kuehn, K.A.; Suberkropp, K. Decomposition of standing litter of the freshwater emergent macrophyte *Juncus effuses. Freshw. Biol.* **1998**, *40*, 717–727. [CrossRef]
80. Sujatha, S.; Bhat, R. Impacts of vermicompost and nitrogen, phosphorus, and potassium application on soil fertility status in arecanut grown on a laterite soil. *Commun. Soil Sci. Plant Anal.* **2012**, *43*, 2400–2412. [CrossRef]
81. Hu, W.; Coomer, T.; Loka, D.; Oosterhuis, D.; Zhou, Z. Potassium deficiency effects the carbon-nitrogen balance in cotton leaves. *Plant Physiol. Biochem.* **2017**, *115*, 408–417. [CrossRef] [PubMed]
82. Dinis, M.J.; Bezerra, R.M.; Nunes, F.; Dias, A.; Guedes, C.; Ferreira, L.M.; Cone, J.W.; Marques, G.; Barros, A.; Rodrigues, M. Modification of wheat straw lignin by solid state fermentation with-rot fungi. *Bioresour. Technol.* **2009**, *100*, 4829–4835. [CrossRef]
83. Osada, M.; Hiyoshi, N.; Sato, O.; Arai, K.; Shirai, M. Effect of sulfur on catalytic gasification of lignin in supercritical water. *Energy Fuels* **2007**, *21*, 1400–1405. [CrossRef]
84. Brijwani, K.; Rigdon, A.; Vadlant, P. Fungal laccase: Production, function, and applications in food processing. *Enzyme Res.* **2010**. [CrossRef]
85. Coronel, L.M.; Joson, L.M.; Mesina, O.G. Isolation and screening of thermophilic fungi for cellulose production. *Philipp. J. Sci.* **1991**, *120*, 379–389.
86. Irbe, I.; Elisashvili, V.; Asatiani, M.; Janberga, A.; Andersone, I.; Andersons, B.; Biziks, V.; Grinins, J. Lignocellulolytic activity of *Coniophora puteana* and *Trametes versicolor* in fermentation of wheat bran and decay of hydrothermally modified hardwoods. *Int. Biodeterior. Biodegrad.* **2014**, *86*, 71–78. [CrossRef]
87. Sukumaran, R.; Singhania, R.; Pandey, A. Microbial cellulases—Production, applications and challenges. *J. Sci. Ind. Res. India* **2005**, *64*, 832–844.
88. Arantes, V.; Saddler, J. Access to cellulose limits the efficiency of enzymatic hydrolysis: The role of amorphogenesis. *Biotechnol. Biofuels* **2010**, *3*, 1–11. [CrossRef] [PubMed]
89. Du, J.; Pu, G.; Shao, C.; Cheng, S.; Cai, J.; Zhou, L.; Jía, Y.; Tian, X. Potencial of extracellular enzymes from *Trametes versicolor* F21a in *Mycrocystis* spp. degradation. *Mater. Sci. Eng. C* **2015**, *48*, 138–144. [CrossRef] [PubMed]
90. Salinas, A.; Vega, M.; Lienqueo, M.; Garcia, A.; Carmona, R.; Salazar, O. Cloning of novel cellulases from cellulolytic fungi: Heterologous expression of a family 5 glycoside hidrolase from *Trametes versicolor* in *Pichia pastoris. Enzyme Microb. Tech.* **2011**, *49*, 485–491. [CrossRef] [PubMed]

 *fermentation*

Article

Spent Yeast from Brewing Processes: A Biodiverse Starting Material for Yeast Extract Production

Friedrich Felix Jacob [1,*], Lisa Striegel [2], Michael Rychlik [2], Mathias Hutzler [3] and Frank-Jürgen Methner [1]

[1] Technische Universität Berlin—Institut für Lebensmitteltechnologie und Lebensmittelchemie—Fachgebiet Brauwesen, Seestraße 13, 13353 Berlin, Germany; frank-juergen.methner@tu-berlin.de

[2] Technische Universität München—Lehrstuhl für Analytische Lebensmittelchemie, Maximus-von-Imhof-Forum 2, 85354 Freising, Germany; lisa.striegel@tum.de (L.S.); michael.rychlik@tum.de (M.R.)

[3] Technische Universität München—Forschungszentrum Weihenstephan für Brau-und Lebensmittelqualität, Alte Akademie 3, 85354 Freising-Weihenstephan, Germany; m.hutzler@tum.de

* Correspondence: f.jacob@campus.tu-berlin.de

Received: 5 June 2019; Accepted: 20 June 2019; Published: 24 June 2019

Abstract: Spent yeast from beer manufacturing is a cost-effective and nutrient-rich starting material for the production of yeast extracts. In this study, it is shown how physiologically important ingredients in a yeast extract are influenced by the composition of the spent yeast from the brewing process. In pilot fermentations, the time of cropping (primary fermentation, lagering) of the spent yeast and the original gravity (12 °P, 16 °P, 20 °P) of the fermentation medium was varied, and four alternative non-*Saccharomyces* yeast strains were compared with two commercial *Saccharomyces* yeast strains. In addition, spent yeast was contaminated with the beer spoiler *Lactobacillus brevis*. The general nutrient composition (total protein, fat, ash) was investigated as well as the proteinogenic amino acid spectrum, the various folate vitamers (5-CH$_3$-H$_4$folate, 5-CHO-H$_4$folate, 10-CHO-PteGlu, H$_4$folate, PteGlu) and the biological activity (reduction, antioxidative potential) of a mechanically (ultrasonic sonotrode) and an autolytically produced yeast extract. All the investigated ingredients from the yeast extract were influenced by the composition of the spent yeast from the brewing process. The biodiversity of the spent yeast from the brewing process therefore directly affects the content of physiologically valuable ingredients of a yeast extract and should be taken into consideration in industrial manufacturing processes.

Keywords: yeast extract; brewer's spent yeast; autolysis; ultrasonic sonotrode; *Saccharomyces cerevisiae/pastorianus*; non-*Saccharomyces* yeast; proteinogenic amino acids; folate vitamers; biological activity

1. Introduction

Beer production generates large quantities of spent yeast during the fermentation and lagering process. Following primary fermentation, this equates to about 0.7–1.1 kg compressed yeast per hectoliterfinished beer [1]. According to the current state of brewing technology, spent yeast after primary fermentation is only used in small quantities to pitch the next batch [2]. The major share is obtained from a propagation plant, which provides highly viable and vital yeast that ferments vigorously [2]. At the end of the cold lagering process, yeast referred to as "lagering cellar yeast" is generated (0.5–0.9 kg compressed yeast per hl finished beer), together with precipitated turbidity particles and "barm beer" [1].

The spent yeast from the brewing process is suitable for use as an efficient starting material to produce yeast extract [1,3]. Yeast extract is generally defined as the soluble content of a yeast cell that remains once the cell wall has been destroyed and removed [4–6]. The variety of different physiologically valuable substances in yeast cells offers the possibiliy of using them as yeast extract in different areas of the food industry [3,7]. As "yeast food", these extracts can therefore increase the free α-amino nitrogen (FAN) when fermenting beer worts with a high content of unmalted grains [8] or a high extract content (high-gravity worts) [9,10], and consequently improve the yeast's nutrient supply and fermentation performance [11,12]. Free proteinogenic amino acids supply the majority of the FAN [12]. The quantity and composition of the relevant amino acids are ultimately critical to performance during fermentation [13] and also impact the beer's aroma profile [14]. From a nutritional standpoint, yeast extracts from spent yeast supply a high concentration of essential and semi-essential amino acids for human beings [5,7]. Yeast extracts are also a good source of B vitamins [6,15]. Among these, the various naturally occurring folate vitamers play an essential role in the human diet, with the biologically active form 5-methyltetrahydrofolate (5-CH$_3$-H$_4$folate) fulfilling key metabolic tasks in human cells [16]. The bioactivity of yeast extracts, which is demonstrated in the form of reduction and anti-oxidative potential, also makes these extracts particularly interesting for the food industry [6,7,15,17].

The majority of globally produced beer is manufactured by fermenting high-gravity worts [18]. The extract content of the wort is increased by adding sugar syrup, which modifies the nutrient balance of the wort with respect to all physiologically active components [18]. The altered yeast metabolism during high-gravity fermentation not only changes the quality of the finished beer but also the material composition of the yeast [10,18]. In addition, the biodiversity of the generated spent yeast in breweries is increased through the use of various alternative non-*Saccharomyces* strains as pure starter cultures for beer production [19,20]. Improper storage or handling of the spent yeast can result in contamination with various microorganisms, which can impact the subsequent yeast extract production process [21]. The composition of ingredients in commercially available yeast extracts varies greatly [17]. One reason is the influence of different yeast extract manufacturing methods, which we have evaluated in previous studies [5,6]. Another reason lies in the diversity of yeast starting material [22]. To the best of our knowledge, no research has been undertaken until now about the influence of the biodiversity of spent yeast from the brewing process on the composition of ingredients in yeast extracts.

In this work, it was shown for the first time how the composition of various physiologically valuable substance groups of a yeast extract depends on the biodiversity of the spent yeast from beer production. Therefore, beer was produced on a pilot scale using 12 °P wort and different yeast strains (*S. cerevisiae* TUM 68, *S. pastorianus* TUM 34/70, *Saccharomycodes ludwigii* TUM SL 17, *Saccharomycopsis fibuligera* TUM 525, *Brettanomyces bruxellensis* TUM Bret 1 and *Torulaspora delbrueckii* T 90). Furthermore, different wort gravities (12 °P, 16 °P, 20 °P) were fermented with *S. cerevisiae* TUM 68 to investigate the influence of high-gravity brewing on a commercial yeast strain. The spent yeast generated after primary fermentation and lagering was then processed into yeast extract using a mechanical (ultrasonic sonotrode) and autolytic cell disruption method. All yeast extracts were investigated to determine their general composition (protein, fat and ash). The physiologically valuable protein content was analyzed in detail with regard to different free and protein-bound amino acids. The effects on the amino acid spectrum of the yeast extract through contamination of the spent yeast by the obligate beer spoiler *Lactobacillus brevis* were also observed. Additionally, we characterized the biological activity of the yeast extract based on its reduction and anti-oxidative potential. We also showed how the total folate content was allocated between the different folate vitamers (5-CH$_3$-H$_4$folate, 5-CHO-H$_4$folate, 10-CHO-PteGlu, H$_4$folate, PteGlu) in the fermentation medium or in the relevant spent yeast and then how it could be transferred to the yeast extract. These results should increase the knowledge on fluctuating nutritional composition of yeast extracts. Furthermore, the most appropriate brewer's spent yeast could be selected to produce a yeast extract with the desired nutritional composition.

2. Materials and Methods

2.1. Yeast Propagation and Fermentation

A sterilized, hopped and standardized all malt wort concentrate (N53940; Döhler GmbH, Darmstadt, Germany) was used to produce the standardized propagation and fermentation wort. The all malt wort concentrate was diluted to an original gravity of 12 °P. To adjust the wort to the higher gravities of 16 °P and 18 °P, respectively, D-(+)-maltose monohydrate (Merck, Darmstadt, Germany) was added to 12 °P wort as an adjunct. Before use, worts were heat treated at 100 °C for 10 minutes for sterilization. The original gravity of the standardized fermentation wort corresponded to the original gravity of the standardized propagation wort. For the precise composition of the standardized all malt wort refer to Table 1.

Table 1. Wort composition.

Parameter	Amount
Original gravity (°P)	12.00
pH	5.17
Spec. weight SL 20/20 °C	1.04
Zinc (mg/L)	0.15
FAN (mg/100 mL)	25.00
Total AS (mg/100 mL)	201.38
Total sugar (g/L)	80.03
EBC-Bittering units (EBU)	20.00
Glucose (g/L)	10.46
Fructose (g/L)	2.17
Sucrose (g/L)	1.02
Maltose (g/L)	49.34
Maltotriose (g/L)	13.79

The propagation procedure described below was used for all the yeast strains in the study. *Saccharomyces cerevisiae* TUM 68 (hereinafter Scer), *Saccharomyces pastorianus* TUM 34/70 (hereinafter Spas), *Saccharomycodes ludwigii* TUM SL17 (hereinafter Slud), *Saccharomycopsis fibuligera* TUM 525 (hereinafter Sfib), *Brettanomyces bruxellensis* TUM Bret1 (hereinafter Bbru) and *Torulaspora delbrueckii* TUM T90 (hereinafter Tdel) were sourced from the Yeast Center at the Weihenstephan Research Center for Brewing and Food Quality (RCW) of the Technical University of Munich (TUM) on agar slant. An inoculation loop of a pure agar slant colony was transferred to 40 mL standardized wort and incubated for 48 h at 20 °C on an orbital shaker (80 rpm). The 40 mL transferred into 400 mL standard wort and incubated again for 48 h at 20 °C on an orbital shaker (80 rpm). This process was repeated from 400 mL to 4 L standardized wort followed by an incubation at 48 h at 20 °C on an orbital shaker (80 rpm).

The fermentation procedure described below was used for all the yeast strains and standardized worts in the study. Standardized laboratory-scale brewing trials were performed using stainless steel vessels of 10 cm diameter × 33 cm height (2.5 L) with 20% headspace and clamped down lids according to Meier-Dörnberg et al. [23]. The propagated yeast was pitched in standardized and aerated (10 mg O_2/L) wort in a Cornelius container (20 L) with a living cell count of 15 million CFU/mL. Each batch was then divided into three fermentation vessels. Fermentation took place at 18 °C and was unpressurized until final attenuation. The viscous primary fermentation yeast was cropped at the bottom of the vessel immediately after the final attenuation. The fermented supernatant was then stored in a carbonated and pressurized state (0.6 bar) for 14 days at 2 °C. Lagering cellar yeast was cropped from the bottom at the end of the cold lagering process again. The fermented supernatant and the spent yeast of each batch was immediately used for analysis and yeast extract production.

2.2. Yeast Pre-Treatment

After the yeast was cropped, it was immediately subjected to three washing processes to remove residual wort components. Each washing process was performed as follows: The viscous cropped spent yeast was diluted with distilled water to 10% dry matter, passed through a yeast sieve (mesh size 0.5 mm), centrifuged (1000 g, 5 min, 18 °C, 500 mL centrifuge tube) and the supernatant was discarded. The sedimented yeast in the centrifuge tube was then resuspended with distilled water for 5 minutes to 10% dry matter and the washing procedure was started afresh. The washed yeast was subsequently collected and diluted to 7% dry matter in distilled water before being fed into the disruption process.

2.3. Yeast Quality Control

The propagation yeast, the spent yeast, the washed spent yeast before disruption and the macerated yeast suspension were only used after passing quality control. Tests for quality control were already described in detail in our previous work [5,6] and had to give negative results for foreign yeasts and microorganisms.

2.4. Yeast Cell Disruption Methods

2.4.1. Ultrasonic Sonotrode

Cell disruption using cavitation was carried out using the ultrasonic homogenizer SONOPLUS HD 3400 (Bandelin). The sonotrode diameter was 25 mm with an operating frequency of 20 kHz. In a stainless steel vessel (400 mL) the process suspension (200 mL) was subjected for 30 minutes to a constant ultrasonic output of 400 W without pulsation. The resulting process heat was removed by means of a glycol-cooling bath to maintain a constant temperature of 7 °C. This disruption process was adopted according to Jacob et al. [5,6].

2.4.2. Standard Autolysis

To autolyze the yeast cells, 200 mL of the yeast suspension was heated in a reaction vessel (400 mL) for 24 h at 50 °C with constant stirring (100 rpm). Sodium chloride (0.086 mol/L) and ethyl acetate (0.051 mol/L) were added at the start of the process. This disruption process was adopted according to Jacob et al. [5,6].

2.4.3. Autolysis with Contamination of *Lactobacillus brevis*

Lactobacillus brevis BLQ 6 (sourced from the RCW) was cultivated in MRS broth medium (Sifin Diagnostics GmbH, Germany) for 3 days at 28 °C and harvested by centrifugation (2500 g, 10 min). The cells were washed with sterile water and centrifuged again. *Lactobacillus brevis* was added to an autolysis process (resulting in a final concentration in the autolysis suspension of 10^6 CFU/mL) that was conducted as described in Section 2.4.2.

2.4.4. Autolysis to Improve γ-aminobutyric Acid (GABA) Production

Process parameters of Masuda et al. were used to improve GABA production during autolysis [24]. Spent yeast (*S. cerevisiae*) after primary fermentation (12 °P) was washed (see Section 2.2) and added at a dry matter content of 7% to a solution containing sterile distilled water, monosodium glutamate (0.060 mol/L) (Merck, Darmstadt, Germany) and D-(+)-glucose monohydrate (0.266 mol/L) (Merck, Darmstadt, Germany). Following this, 200 mL of the reaction solution was adjusted to pH 6 with 2N HCl or 2N NaOH and incubated at 37 °C for 72 h with constant stirring (100 rpm). After 72 h the reaction solution was heated for 15 min at 85 °C. For the control, the process was conducted without monosodium glutamate or D-(+)-glucose monohydrate.

2.5. Production of Yeast Extract

After the yeast cell disruption process (Sections 2.4.1–2.4.4) cell wall components first had to be separated from the cell extract. Therefore, the samples were centrifuged for 20 minutes at 10,000 g and 4 °C. The supernatant was carefully pipetted out the centrifuge tubes and freeze-dried (Christ Alpha 1–4 LSCbasic, condenser temperature: −55 °C, vacuum: 0.1 mbar, ice condenser capacity: 4 kg/24 h). In this way, a yeast extract powder was produced for the subsequent analyses that offered a constant basis for comparison.

The dry yeast extracts enabled the disruption methods to be directly compared in terms of the following analysis without the need to consider the effectiveness of the different methods. Results on the effectiveness of the three disruption methods can be reviewed in a previous work by Jacob et al. [5]. An overview of the sample description and related process details can be seen in Table 2.

Table 2. Sample overview and process details of cropped surplus yeast after primary fermentation (F.) and cold lagering (L.); data are expressed as mean values; confidence limits were determined to be lower than 5% of the average value.

Sample Name	Yeast Species	Yeast Strain	Original Gravity (°P)	Apparent Attenuation (%)	Yeast Crop After
Scer 12°P	*Saccharomyces cerevisiae*	TUM 68	12	79	F.
Scer 12°P L	*Saccharomyces cerevisiae*	TUM 68	12	79	L.
Scer 16°P	*Saccharomyces cerevisiae*	TUM 68	16	85	F.
Scer 20°P	*Saccharomyces cerevisiae*	TUM 68	20	89	F.
Spas 12°P	*Saccharomyces pastorianus*	TUM 34/70	12	80	F.
Slud 12°P	*Saccharomycodes ludwigii*	TUM SL 17	12	11	F.
Tdel 12°P	*Torulaspora delbrueckii*	TUM T 90	12	45	F.
Bbru 12°P	*Brettanomyces bruxellensis*	TUM Bretta 1	12	40	F.
Sfib 12°P	*Saccharomycopsis fibuligera*	TUM 525	12	60	F.

2.6. Analysis

2.6.1. Protein and Amino Acids

Nitrogen content in the yeast extract was determined using the Kjeldahl method described in the MEBAK (Central European Brewing Technology Analysis Commission) brewing technology analysis methods (Method 2.6.1.1) [25]. Protein content of the yeast extracts was estimated by multiplying its nitrogen content by the factor 5.5 [26].

Free proteinogenic amino acids (except proline and cysteine) were quantified using high performance liquid chromatography (HPLC) according to MEBAK Method 2.6.4.1 [27]. The detailed procedure for proline and cysteine (Method 4.11.1) was taken from *Buch für chemische Untersuchung von Futtermitteln* (The chemical analysis of feedstuffs) [25]. To determine all free and protein-bound amino acids (total amino acid quantity), the resuspended yeast extracts underwent acid hydrolysis before measurements were taken according to Method 4.11.1 of the chemical analysis of feedstuffs [25].

2.6.2. Fat

Crude fat was determined according to Method 5.1.1 from the *Methodenbuch für chemische Untersuchung von Futtermitteln* (The chemical analysis of feedstuffs) [25].

2.6.3. Water and Ash Content

Water content was determined using MEBAK Method 2.2 [27], the ash content similarly according to Method 8.1 from the *Methodenbuch für chemische Untersuchung von Futtermitteln* (The chemical analysis of feedstuffs) [25].

2.6.4. Folate

Vitamin B9 (total folate) and folate derivatives (5-CH$_3$-H$_4$folate, 5-CHO-H$_4$folate, 10-CHO-PteGlu, H$_4$folate, PteGlu) were analyzed using LC-MS/MS according to the recently published method by Striegel et al. [28].

2.6.5. Antioxidative and Reduction Potential

The antioxidative potential of yeast extracts was measured using a Sigma-Aldrich kit, in which the antioxidants from a sample inhibit the formation of radical cations. Spectrophotometry was used to measure this inhibition proportionally by means of a color reaction. Trolox (TE), a vitamin E analog, was used as the control antioxidant.

Reduction potential of the yeast extract can be determined by MEBAK Method 2.15.2 [27]. The reductones of the sample reduce a certain quantity of Tillmann's reagent (2,6-dichlorophenolindophenol, DPI) within a certain period, which can be measured spectrophotometrically (520 nm).

2.6.6. Wort Density

Wort density was measured using a DMA™ 35 Basic portable density meter (Anton Paar GmbH, Ostfildern, Germany), and the gravity was expressed in degree Plato (°P). 1 °P corresponded to 1 g of extract per 100 g of liquid solution, where extract included both fermentable sugars and non-fermentable carbon sources.

2.6.7. Calculations of Apparent Attenuation

Apparent attenuation (%) of the wort was the proportion of the wort dissolved solids (extract), which was fermented during fermentation:

Apparent attenuation (%) = [(original gravity (°P) − final gravity (°P))/(original gravity (°P))] × 100.

2.7. Statistical Evaluation

All experiments were performed in triplicate and the relevant results given as arithmetic means. At a confidence level of 95%, the expected range (confidence interval) for each mean was calculated from the variance using Student's t-distribution. A single factor analysis of variance (ANOVA) and a paired t-test were performed to demonstrate differences between the results. "Significant" differences were described as having a p-value < 0.05. A test by Dixon was used to evaluate the results.

3. Results and Discussion

3.1. General Nutrient Composition

The nutrient value of the generated yeast extracts showed great variability. Table 3 lists the analysis results of the general nutrient composition of yeast extracts produced using the mechanical (ultrasonic sonotrode) method. To calculate the total protein content of the yeast extracts, the nitrogen quantity determined via Kjeldahl analysis was multiplied by the conversion factor 5.5 proposed by Reed et al. [26] This was proven to be a suitable conversion factor in our previous published papers in relation to yeast extracts [5,6] and was also used by Caballero-Cordoba et al. [29]. The factor of 6.25, which is generally used, overestimates the protein content as the total nitrogen volume contains the RNA nitrogen quantity (ribonucleic acids, 5–10% of the dry mass of yeast extract) as well as the proteinogenic nitrogen [5,6]. In the statistical evaluation of the obtained results, we established that the total protein content of the mechanically produced yeast extracts did not differ significantly (ANOVA p-value > 0.05) from the protein content of the autolytically produced yeast extracts (all nutritional data of the autolytically produced yeast extracts (Table S1) can be found in the supplementary materials). This observation was already noted and discussed in one of our previous papers [6]. Yeast extracts produced using primary fermentation yeast with different original wort contents (Scer 12 °P, Scer

16 °P, Scer 20 °P), differ significantly (ANOVA *p*-value < 0.05), with no significant difference between the fermentations Scer 16 °P and Scer 20 °P (*t*-test *p*-value > 0.05) (Table 3). This fact is justified in that adding maltose to the high-gravity worts (Scer 16 °P, Scer 20 °P) modified the nutrient balance compared with the normal-gravity worts (Scer 12 °P), and this caused a higher osmotic pressure at the start of fermentation and a higher alcohol content at the end of fermentation. Consequently, the yeast's vitality and viability dropped [30], which is associated with reduced specific growth and fermentation rates [18,31]. For high-gravity fermentations, this is also linked to reduced amino acid uptake rates, higher residual FAN (freely available amino nitrogen) and increased accumulation of trehalose and glycogen [1,18]. Presumably, this reduces the protein content of the yeast cell dry mass and ultimately results in a lower total protein content in the yeast extract. In our trials, when using spent yeast from high-gravity fermentations (Scer 16 °P, Scer 20 °P), we established a protein content in yeast extract reduced by 8.5% when compared with spent yeast from normal-gravity (Scer 12 °P) fermentations.

Table 3. General nutritional composition of yeast extracts made from spent yeast of beer production via a mechanical disruption method (ultrasonic sonotrode); influence of original gravity (12 °P, 16 °P, 20 °P), time of yeast cropping (after primary fermentation vs. after cold beer lagering (L)) and yeast strain (Scer, Spas, Slud, Tdel, Bbru, Sfib); for results of ANOVA and pairwise t-test, see text; data are expressed as mean values ± confidence limits; Cal.: calculated sum parameter of carbohydrates, RNA, non-nitrogen fraction and others components.

Sample Name	Scer 12°P	Scer 12°P L	Scer 16°P	Scer 20°P	Spas 12°P	Slud 12°P	Tdel 12°P	Bbru 12°P	Sfib 12°P
Protein (mg/g dw) (N × 5.5)	480.08 ± 10.25	448.53 ± 4.93	411.80 ± 9.61	395,26 ± 5.67	411.54 ± 8.62	574.28 ± 11.23	446.81 ± 8.57	508.34 ± 8.75	598.01 ± 13.88
Fat (mg/g dw)	10.6 ± 0.32	10.5 ± 0.31	10.3 ± 0.36	10.4 ± 0,52	11.8 ± 0.72	15.5 ± 0.43	9.88 ± 0.53	18.2 ± 0.37	9. 68 ± 0.69
Ash (mg/g dw)	130.55 ± 2.49	120.11 ± 3.45	84.20 ± 2.73	87.10 ± 3.02	82.10 ± 1.98	110.40 ± 2.57	79.12 ± 1.94	130.64 ± 2.47	89.70 ± 3.24
Cal. (mg/g dw)	378.77	420.86	493.70	507.24	494.56	299.82	464.19	342.82	302.61
Moisture (%)	9.30 ± 0.11	9.35 ± 0.13	9.41 ± 0.10	9.46 ± 0.12	9.34 ± 0.13	9.43 ± 0.12	9.31 ± 0.13	9.32 ± 0.11	9.29 ± 0.14

We could also see a significant reduction (*t*-test *p*-value < 0.05) of the protein content (3%) in the yeast extract when using lagering cellar yeast (Scer 12 °P L) compared with primary fermentation yeast (Scer 12 °P) (Table 3). At the end of primary fermentation, part of the yeast was still suspended and settled as "lagering cellar yeast" only once cold lagering began. In this connection, Powell et al. showed that the unsettled yeast had a lower cell age with reduced fermentation performance and flocculation tendency, suggesting a modified physiological state [32]. A modified physiological state could explain the different protein contents of the yeast extract produced from lagering cellar yeast (Scer 12 °P L) and primary fermentation yeast (Scer 12 °P), too. During longer lagering phases, proteinogenic material in the yeast could also be lost via excretion, as established by Steckley et al. [33].

The protein content of the yeast extracts produced from primary fermentation yeast with different yeast strains also differed significantly (ANOVA *p*-value < 0.05). The non-*Saccharomyces* yeast strain *S. fibuligera* TUM 525 (Sfib 12 °P) gave the highest value at 598 mg/g yeast extract. The strain often used to produce alcohol-free or low-alcohol beer, *S. ludwigii* TUM SL 17 (Slud 12 °P), had the second-highest value at 574 mg/g yeast extract. The yeast strain *B. bruxellensis* TUM Bret 1 (Bbru 12 °P) provided 508 mg protein per g yeast extract. Only for the non-*Saccharomyces* yeast strain *T. delbrueckii* TUM T 90 (Tdel 12 °P) and the two commercially used yeast strains *S. cerevisiae* TUM 68 (Scer 12 °P) or *S. pastorianus* TUM 34/70 (Spas 12 °P) the protein content was less than 500 mg/g yeast extract. Spas 12 °P gave the lowest value overall in our trials (411.54 mg/g). For a *S. pastorianus* yeast strain, Vieira et al. determined values of 698 mg/g and 765 mg/g yeast extract when reusing the yeast two to four times in the fermentation process [34]. In another work, the same group determined a protein content of 641 mg/g yeast extract [15]. The sampling time during fermentation was unknown and the higher conversion factor of 6.25 was applied to calculate the protein content. For a brewery spent yeast,

Podpora et al. established a protein content of 625 mg/g or 638 mg/g yeast extract, without giving details on the yeast strain or process conditions. Protein from the fermentation medium was also recorded for the yeast extract production and the higher conversion factor of 6.25 was applied [35].

There was no significant difference in the ash content of mechanically and autolytically produced yeast extracts (ANOVA p-value > 0.05), as demonstrated in our previous work when investigating various disruption methods [6]. The use of spent yeast from high-gravity fermentations (Scer 16 °P, Scer 20 °P) gave significantly lower values (t-test p-value < 0.05) for the ash content in the yeast extract than a spent yeast from normal-gravity fermentation (Scer 12 °P). This observation can be presumably attributed to the same effect that reduced the protein content in the yeast extract. Therefore, the ash percentage of the total dry mass could presumably be reduced by an increased trehalose and glycogen content due to the modified yeast metabolism. Another significantly lower result (t-test p-value < 0.05) was the ash concentration in the yeast extract caused by using lagering cellar yeast (Scer 12 °P L) rather than spent yeast obtained following primary fermentation (Scer 12 °P). The different physiological state of these two starting yeasts mentioned earlier is also suspected of influencing the cell ingredient composition. The ash content of the yeast extracts of all investigated yeast strains differed significantly (ANOVA p-value < 0.05). Only Bbru 12 °P and Scer 12 °P showed no significant difference (t-test p-value > 0.05). In literature, for the ash content, a range from 78 mg/g to 140 mg/g yeast extract is found [15,34,35]. Due to different yeast strains, fermentation media and yeast extract production processes, a direct comparison is not possible.

The fat content of all autolytically produced yeast extracts was between 0.04–0.05% of the dry mass and did not differ significantly (ANOVA p-value > 0.05). By using the mechanical cell disruption method, the fat content of the yeast extracts reached a maximum of 18.2 mg and a minimum of 9.68 mg per g yeast extract (Table 3). Neither the time of cropping the spent yeast (Scer 12 °P vs. Scer 12 °P L) or the original wort content (Scer 12 °P, Scer 16 °P, Scer 20 °P) in the fermentation process had any significant influence on the fat content of the yeast extract (ANOVA and t-test p-value > 0.05). Only the yeast strains Slud 17 12 °P and Bbru 12 °P differed significantly from the others (t-test p-value < 0.05). In general, the fat content of the yeast extracts was very low as already established in other studies [6,15,34].

3.2. Amino Acid Composition

From a physiological point of view, the proteinogenic material composition is crucial. The essential amino acids (His, Thr, Val, Met, Ile, Phe, Leu Lys) are indispensable for human nutrition as these cannot be synthesized by the body and must be absorbed via food [36]. The rapid usability of proteinogenic material in microbiological culture media is especially assured for yeasts if this material is present in the form of free amino acids, i.e. individual amino acids are not linked via peptide bonds [13]. This ensures the amino acids can be transported via various mechanisms through the cell wall and cell membrane and then to be metabolized [13]. Specific amino acids are preferentially absorbed by the cell [13]. In addition, individual amino acids can significantly affect the aroma metabolism of a yeast and thereby influence the overall aroma of a fermentation by-product [14]. In a previous study, we showed how different cell disruption methods impacted the amino acid composition and the proteinogenic material [5]. Following an autolytic process, the content of free amino acids in the yeast extract was significantly higher than that produced via mechanical disruption methods due to enzymatic degradation processes [5]. This was also confirmed in the trials presented here, as revealed by the comparison between mechanical (Figure 1) and autolytic (Figure 2) methods for the relevant trial series (t-test p-value < 0.05).

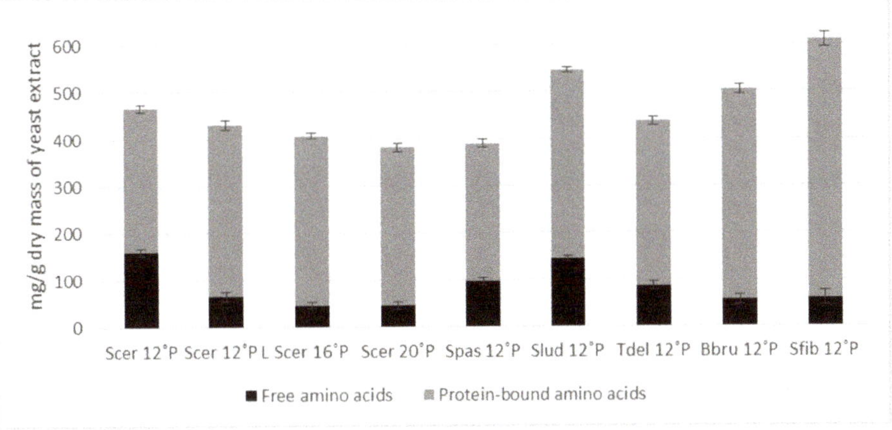

Figure 1. Free and protein-bound amino acids in yeast extracts made from spent yeast of beer production via mechanical disruption method (ultrasonic sonotrode); influence of original gravity (12 °P, 16 °P, 20 °P), time of yeast cropping (after primary fermentation vs. after cold beer lagering (L)) and yeast strain (Scer, Spas, Slud, Tdel, Bbru, Sfib); for results of ANOVA and pairwise t test, see text; data are expressed as mean values ± confidence limits.

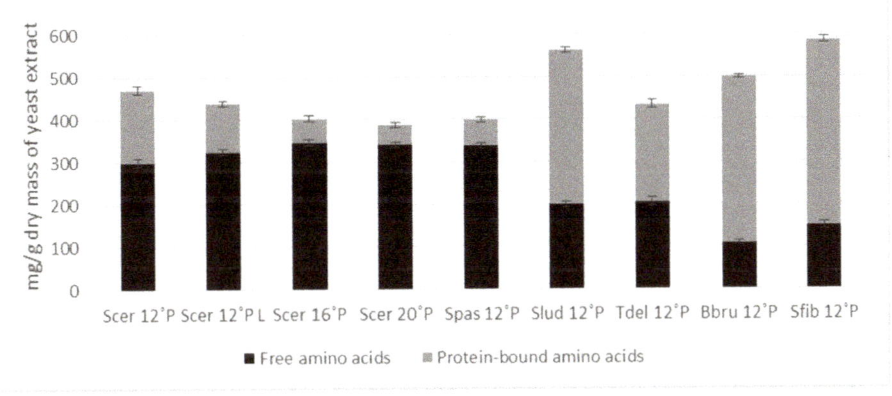

Figure 2. Free and protein-bound amino acids in yeast extracts made from spent yeast of beer production via autolysis; influence of original gravity (12 °P, 16 °P, 20 °P), time of yeast cropping (after primary fermentation vs. after cold beer lagering (L)) and yeast strain (Scer, Spas, Slud, Tdel, Bbru, Sfib); for results of ANOVA and pairwise *t* test, see text; data are expressed as mean values ± confidence limits.

However, the aim of this work was to evaluate the influence of the starting material on the amino acid composition of the yeast extract according to the two disruption methods. It was shown that producing the yeast extract via sonotrode (Figure 1) generated significant differences (ANOVA *p*-value < 0.05) in the content of free amino acids in the yeast extract when using spent yeast from fermentation processes with different gravities (Scer 12 °P, Scer 16 °P, Scer 20 °P). No significant difference was recorded between the Scer 16 °P and Scer 20 °P test series (*t*-test *p*-value > 0.05). Using lagering cellar yeast (Scer 12 °P L) also generated a significantly lower content of free amino acids (*t*-test *p*-value < 0.05). In the comparison of the test series Scer 12 °P, Scer 12 °P L, Scer 16 °P, Scer 20 °P, it was also apparent that lower total quantities of all amino acids resulted in lower quantities of free amino acids. The content of free amino acids following mechanical disruption was likely to be derived largely from the free amino acid pool in the cell [1,5], which is influenced by the extraction process [37]. Amino acids that were enzymatically released from the protein (despite a constant process temperature of 7 °C), were also present in the mechanically produced yeast extract as already evidenced in a previous paper [5]. The content of free amino acids in the yeast extract of yeast strains Scer 12 °P, Spas 12 °P, Slud 12 °P, Sfib 12 °P, Bbru 12 °P and Tdel 12 °P were significantly different (ANOVA *p*-value < 0.05 (Figure 1). There was no correlation between the total quantity of all amino acids (or bound amino acids) and the free amino acids in the yeast extract.

Significant differences were established for the test series Scer 12 °P and Scer 12 °P L (*t*-test *p*-value < 0.05) or Scer 12 °P, Scer 16 °P and Scer 20 °P (ANOVA *p*-value < 0.05) and no difference between Scer 16 °P and Scer 20 °P (*t*-test *p*-value > 0.05) following the autolytic process (Figure 2). When comparing the test series Scer 12 °P, Scer 12 °P L, Scer 16 °P, Scer 20 °P, it was also shown that the lower the total quantity of all amino acids, the higher the quantity of free amino acids. This indicated that the higher gravity in the fermentation process or due to the longer lagering period the spent yeast obtained during the autolytic yeast extract production process had a greater hydrolytic potential. The increased release of amino acids from the cell protein during the autolytic process was presumably attributed to a higher quantity of various proteinases in the yeast cells. An increased amount of proteinases in the fermentation medium was therefore reported for high-gravity fermentations, which is caused by excretion from the living yeast cell as well as by cell autolysis [38,39]. Fukal et al. also reported that yeast proteinases have high thermostability at a temperature of 50 °C [40], which corresponds to the autolysis process temperature selected in this study. It could also be evidenced that low yeast vitality is associated with greater proteinase excretion [38]. Yeast vitality at the end of lagering also drops substantially with beer production, whereby proteinase is released [38]. This could explain the increased proteolytic activity of the lagering cellar yeast during autolytic yeast extract production. Consequently, adding spent yeast from high-gravity fermentations and using lagering cellar yeast results in a higher amount of free amino acids in the yeast extract. From Figure 3 it is obvious that individual amino acids in the relevant test series (Scer 12 °P, Scer 12 °P L, Scer 16 °P, Scer 20 °P) are also released from the protein in different percentages. This means that not only the total quantity of free amino acids differed between the individual test series, but also the spectrum of individual free amino acids in the yeast extract. Yeast cells contain a variety of different proteolytic enzymes [41,42], which are likely to be present and active in different quantities in the respective test series. The range of individual free amino acids in the yeast extract therefore depends not only on the production method, as already shown [5], but also on the original wort content and the lagering period of the beer, from which the spent yeast originates. For the different yeast strains Scer 12 °P, Spas 12 °P, Slud 12 °P, Sfib 12 °P, Bbru 12 °P and Tdel 12 °P, we could show a significant difference (ANOVA *p*-value < 0.05) in the content of free amino acids in the autolytically produced yeast extracts (Figure 2). However, there was no correlation between the total quantity of amino acids and the free amino acids. The yeast extracts from spent yeast of non-*Saccharomyces* yeast contained a maximum of 200 mg free amino acids per g yeast extract. In contrast, the two commercially used yeast strains Scer 12 °P and Spas 12 °P provided 340 mg/g. Berlowska et al. determined 449.7 mg free amino acids per g yeast extract for a *S. cerevisiae* yeast strain [22]. For the analyzed non-*Saccharomyces* yeast species *K. marxianus, S. stipitis* and *P. angusta*, a free amino acid content was ranging of between 101.4 mg and 405.3 mg per g yeast extract [22]. It is not possible to directly compare these results with the current study due to the different production process and fermentation conditions. In the exemplary comparison, Figure 4 presents the detailed spectrum of free amino acids of yeast extracts, produced from spent yeast of the commercial yeast strain Spas 12 °P and the alternative non-*Saccharomyces* yeast strain Slud 12 °P. Significant differences (*t*-test *p*-value < 0.05) were found for the amino acids Asp, Glu, Asn, Ser, Gly, Thr, Tyr, Val, Trp, Ile, Phe and Leu. The amount of all individual amino acids (total, free) of conducted experiments can be found in Figures S1–S4 of the supplementary materials.

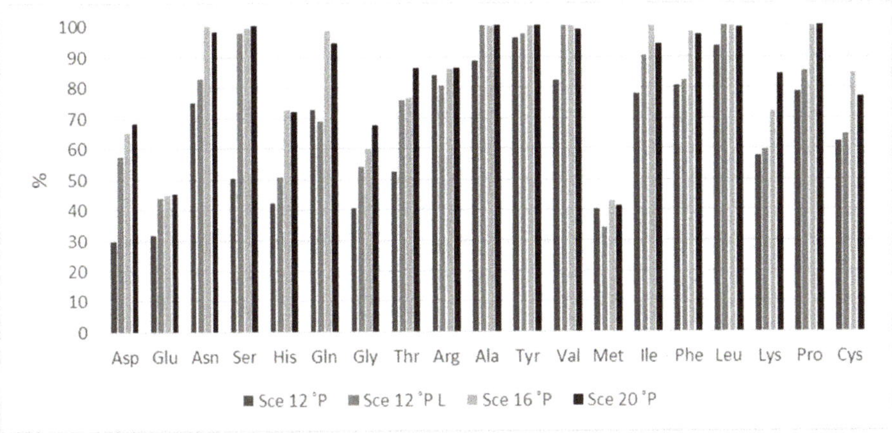

Figure 3. Percentage share of free amino acids of yeast extract (from *S. cerevisiae* TUM 68) protein released via autolysis; influence of original gravity (12 °P, 16 °P, 18 °P) and time of yeast cropping (after primary fermentation vs. after cold beer lagering (L)).

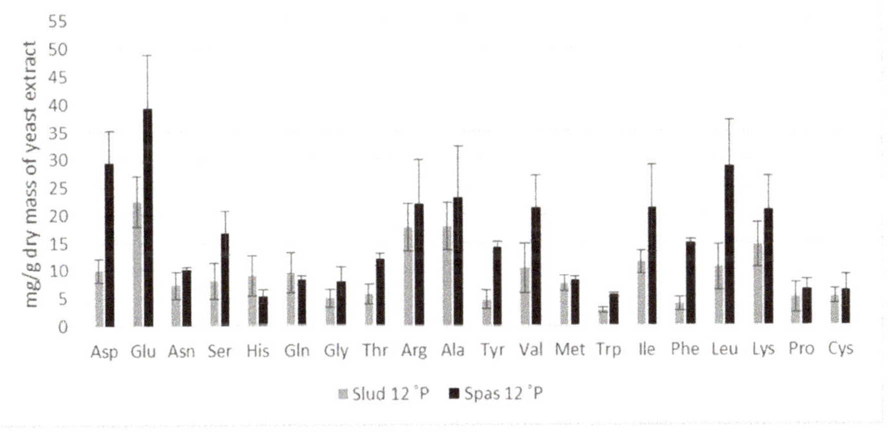

Figure 4. Free proteinogenic amino acid spectrum in yeast extracts made from spent yeast (*S. pastorianus* TUM 34/70 and *S. ludwigii* TUM SL 17) after primary fermentation via autolysis; for results of ANOVA and pairwise t-test, see text; data are expressed as mean values ± confidence limits.

The proteinogenic amino acid composition of the yeast extract depends on the production method [5] and, as shown above, on the starting yeast. If hygiene standards are not maintained when producing beer and storing yeast, the starting yeast could potentially be contaminated with microorganisms. The most common beer-spoilage organism in early stages of the beer production process is the species *Lactobacillus brevis* [43]. The ability of this bacteria species to convert glutamic acid into the nutritionally valuable γ-aminobutyric acid (GABA) has been proven [44]. Masuda et al. demonstrated that this reaction also proceeds during the autolysis of different yeast strains [24]. An intrinsic enzymatic mechanism with the enzyme glutamate decarboxylase is responsible for this reaction [24]. An excess of glutamic acid as a substrate and glucose as an energy supplier can increase GABA formation [24]. In a previous study we could establish an increased GABA concentration in the yeast extract for the strain *S. cerevisiae* TUM 68 [5] and also presumed this was caused by the mechanism postulated by Masuda et al. [24]. As all other yeast strains in this study had lower

GABA concentrations (see supplementary material) in their autolytically produced yeast extracts than *S. cerevisiae* TUM 68 (Scer 12 °P, 50 °C), we attempted to further increase GABA formation for the autolysis of *S. cerevisiae* TUM 68 (Scer 12 °P + Gluc + Glu, 37 °C), using the same process parameters as Masuda et al. At the same time, we investigated whether contamination of the yeast with *L. brevis* (Scer 12 °P + L, 50 °C) also produced an elevated GABA concentration in the autolytically produced yeast extract, or altered the proteinogenic amino acid composition. However, contamination with *L. brevis* (Scer 12 °P + L, 50 °C) influenced neither the proteinogenic amino acid spectrum of the yeast extract, nor did the GABA concentration differ significantly from the control sample (Figure 5) (*t*-test *p*-value > 0.05). The results coincided with the data from our previous study [5]. Champagne et al. also reported that they observed no significant influence of bacterial contamination on the extract yield, the total nitrogen, the FAN or the turbidity of an autolytically produced yeast extract [45]. The solvent ethyl acetate presumably inhibited the contaminants, reported by Champagne et al. [45] and in this study. Barrette et al. could therefore show a reduced viability in a bacteria population during autolytic yeast extract production using ethyl acetate [21]. Under the autolysis conditions of Masuda et al., the GABA concentration could be increased in the yeast extract of *S. cerevisiae* (Figure 5), which suggests an enzymatic mechanism with the enzyme glutamate decarboxylase being active. While the GABA concentration in our yeast extract of *S. cerevisiae* roughly doubled, Masuda et al. was able to increase the values for various *Candida* and *Pichia* strains by more than tenfold [24].

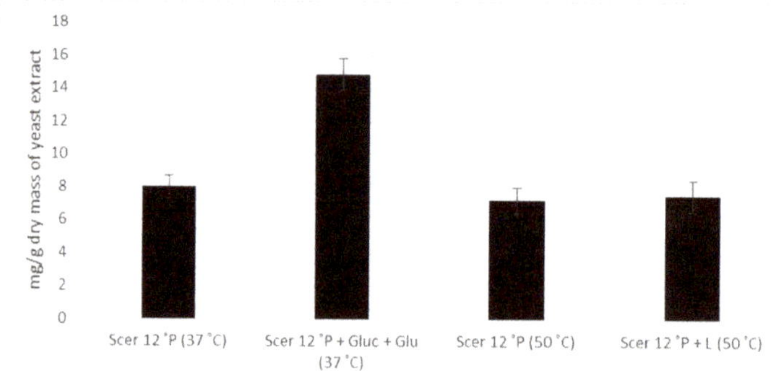

Figure 5. Gamma-aminobutyric acid in yeast extracts made from surplus yeast (*S. cerevisiae* TUM 68) after primary fermentation via autolysis (37 °C, 72 h or 50 °C, 24 h); Scer 12 °P (control); Scer 12 °P + Gluc + Glu (Scer 12 °P + glucose + glutamic acid); Scer 12 °P + L (Scer 12 °P + *Lactobacillus brevis*); for results of ANOVA and pairwise t-test, see text; data are expressed as mean values ± confidence limits.

3.3. Folate Vitamer Distribution

Yeasts and yeast extracts are known for their high content of different B vitamins, which include folates (vitamin B9) [6,15]. Folates, especially tetrahydrofolate-(H_4folate)-polyglutamates, play an essential role in various metabolic pathways such as amino acid synthesis in mitochondria and DNA replication in the cell nucleus [28]. The human body cannot generate vitamin B9 itself and therefore needs to obtain an adequate supply via diet [28]. In a previous study, we already showed that yeast extracts can be produced with a folate content between 1.35 and 4.94 mg/100 g using a spent yeast (*S. cerevisiae* TUM 68) from a top-fermenting brewing process [6]. In this context, we presented the influence of different mechanical and autolytic extraction methods on the content of the folate vitamers 5-CH_3-H_4folate, 5-CHO-H_4folate, 10-CHO-PteGlu, H_4folate, and PteGlu. Hjortmo et al. evidenced a total folate content in a range of 4000–1,4500 µg/100 g yeast dry matter for various yeast strains, wherein the sample material came from the exponential growth phase of the yeast population and a synthetic culture medium was used [46]. The spent yeast from beer preparation is cropped at the end of the fermentation process or as lagering cellar yeast and as such is already in the stationary phase of yeast cell growth. In this phase, the folate content of the yeast dry mass drops sharply and remains at a lower level than in the exponential growth phase [47]. The material composition of the culture medium also plays a critical role [47].

In order to provide a comprehensive picture of the process, we also established the folate content of the fermented medium and the spent yeast in addition to the autolytically and mechanically produced yeast extract. This showed that all fermented media had a low total folate content (6–17 µg/100 g), calculated from the folate vitamers 5-CH$_3$-H$_4$folate, 5-CHO-H$_4$folate, 10-CHO-PteGlu, H$_4$folate and PteGlu. No significant difference could be determined between the investigated fermented media of any test series (Scer 12 °P, Scer 12 °P L, Scer 16 °P, Scer 20 °P, Bbru 12 °P, Slud 12 °P, Tdel 12 °P, Spas 12 °P, Sfib 12 °P) (ANOVA p-value > 0.05). Hjortmo et al. could not show any folate in the fermented medium of a *S. cerevisiae* strain [47]. This can be explained by the fact that there were no more yeast cells in the supernatant [47]. In our work, however, the supernatants still contained suspended yeast cells, which were not removed by centrifuging before folate analysis. Regarding the yeast extract, the original wort content (Scer 12 °P, Scer 16 °P, Scer 20 °P) had no significant influence (ANOVA p-value > 0.05) on the total folate content of the spent yeast (Figure 6). High-gravity fermentations generally lead to lower specific yeast growth rates than normal-gravity fermentations [18]. For a continuous yeast culture (chemostat), Hjortmo et al. demonstrated a significant positive correlation between specific growth and the total folate content of a yeast population [47]. This has been evidenced for the exponential growth phase of a yeast population [47]. The primary fermentation yeast is in the stationary phase. In this case, the total folate content of the spent yeasts from the normal and high-gravity fermentations did not differ and presumably settled at around the same level, irrespective of the effect of the different specific growth rates experienced previously. In contrast, the precipitated lagering cellar yeast (Scer 12 °P L) had a significantly lower total folate content than that of the primary fermentation yeast (t-test p-value < 0.05). During the cold lagering process, the non-precipitated primary fermentation yeast stays in the stationary phase and precipitates within this period as lagering cellar yeast (12 °P L). No more cell division occurs during this period, which presumably caused the constant degradation of the folate vitamers required for amino acid synthesis and DNA replication. The ratio of the individual folate vitamers in the spent yeast (Scer 12 °P, Scer 12 °P L, Scer 16 °P, Scer 20 °P) did not change significantly (ANOVA p-value > 0.05). The yeast extracts produced via sonotrode had higher folate contents in each case than the corresponding spent yeasts. Removing the insoluble cell components (mainly cell walls) enriched the folate content. Once again, no significant influence (ANOVA p-value > 0.05) of the original wort content (Scer 12 °P, Scer 16 °P, Scer 20 °P) on the total folate content or the distribution of the individual folate vitamers could be observed. The total folate content of the autolytically produced yeast extracts was between 800 and 1400 µg/100 g and did not differ significantly (ANOVA p-value > 0.05).

In a comparison of the different yeast strains (Figure 7), Scer 12 °P with 3640 µg/100 g yeast dry mass had the highest total folate content in the spent yeast. A value of 2970 µg/100 g could be determined for the spent yeast of the yeast strain Bbru 12 °P. We established the lowest total folate content (1930 µg/100 g) for the yeast strain Spas 12 °P and could determine no significant difference to Sfib 12 °P (t-test p-value > 0.05). Between the strains Slud 12 °P and Tdel 12 °P there was also no significant difference in the total folate content of the spent yeast (t-test p-value > 0.05) (Figure 7). The quantity of the physiologically valuable folate vitamer 5-CH$_3$-H$_4$folate varied in the spent yeast of all the investigated yeast strains (ANOVA p-value < 0.05). It was also observed that the ratio of 5-CH$_3$-H$_4$folate to the other folate vitamers, 5-CHO-H$_4$folate, 10-CHO-PteGlu, H$_4$folate and PteGlu, differed significantly in the spent yeast of the particular yeast strains (ANOVA p-value < 0.05) (Figure 7). At 2200 µg/100 g the proportion of 5-CH$_3$-H$_4$folate was approximately 60% of the total folate content of the spent yeast Scer 12 °P, whereas this was only around 20% for Tdel 12 °P. The total folate content of the yeast extracts (sonotrode) from the spent yeasts Scer 12 °P, Bbru 12 °P and Spas 12 °P differed significantly from the other mechanically produced yeast extracts (t-test p-value < 0.05). The yeast extracts Slud 12 °P, Tdel 12 °P and Sfib 12 °P, however, did not differ significantly (t-test p-value > 0.05). The high total folate content (6000 µg/100 g) of the yeast extract from the spent yeast Bbru 12 °P was striking with regard to the total folate content of the corresponding spent yeast (2970 µg/100 g). The proportion of the physiologically valuable folate vitamers 5-CH$_3$-H$_4$folate of the total folate constant was also not constant in the mechanically produced yeast extracts. The total folate content of the autolytically produced yeast extracts in Figure 7 was between 800 and 2100 µg/100 g. Only the Tdel 12 °P yeast extract differed significantly (*t*-test *p*-value < 0.05) from the others. The influence of the different production methods (autolysis, sonotrode) on the total folate content and the folate vitamer distribution has already been discussed in our previous work [6].

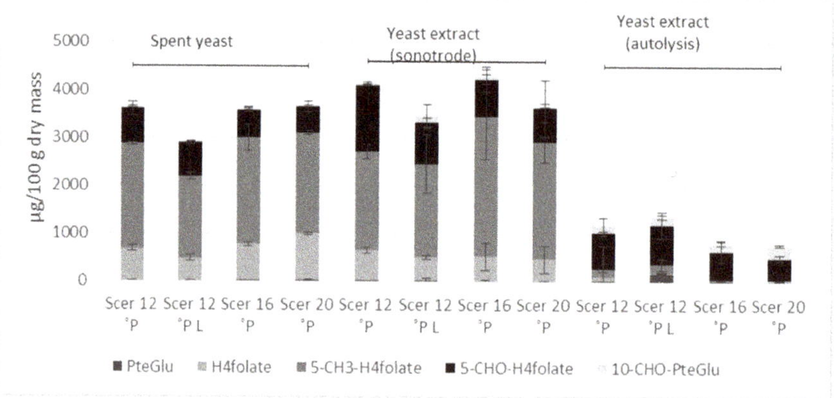

Figure 6. Distribution of the folate vitamers 5-CH$_3$-H$_4$folate, 5-CHO-H$_4$folate, 10-CHOPteGlu, H$_4$folate and PteGlu in spent yeast (*S. cerevisiae* TUM 68) and in the corresponding yeast extract (via sonotrode or autolysis); influence of original gravity (12 °P, 16 °P, 20 °P), time of yeast cropping (after primary fermentation vs. after cold beer lagering (L)); for results of the pairwise t-test, see text; data are expressed as mean values ± confidence limits.

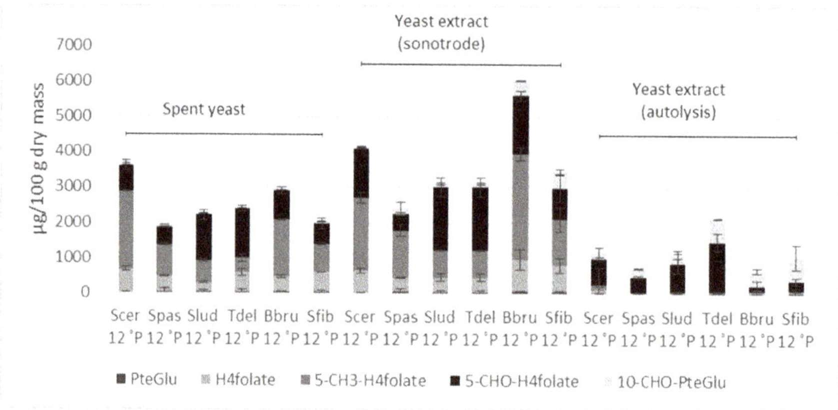

Figure 7. Distribution of the folate vitamers 5-CH$_3$-H$_4$folate, 5-CHO-H$_4$folate, 10-CHOPteGlu, H$_4$folate and PteGlu in spent yeast, yeast extracts (via sonotrode or autolysis) ; influence of yeast strain (Scer, Spas, Slud, Tdel, Bbru, Sfib); for results of the pairwise t-test, see text; data are expressed as mean values ± confidence limits.

3.4. Biological Activity

Yeast cells have a wide range of different functional components, which impart bioactive properties to a yeast extract once extracted from the cell. These include peptides, amino acids, flavonoids, polyphenols and carotenoids [15,34,48]. Through autolysis, Vieira et al. increased the anti-oxidative potential of a mechanically produced yeast extract due to the release of phenolic components and amino acids [49]. When the thermal load is too high, the anti-oxidative potential drops, which, according to Vieira et al., is likely attributed to the breakdown of phenolic components, vitamins and bioactive peptides [49]. We evidenced this correlation in an earlier work as follows. We observed a significant drop in concentration of the bioactive peptide glutathione or the total polyphenol content following an autolytic yeast extract production method exposed to thermal stress compared with a mechanical process at an appropriate thermal load [6]. The process parameters of the two yeast extract production methods from our previous study [6] corresponded to those in these trials. Since the yeast extract of the autolytic process exhibited low bioactivity in the previous study, we only investigated the reductive and anti-oxidative potential of the mechanically produced yeast extract in this work (Figure 8). We found no significant difference in the reduction potential of the test series Scer 12 °P and Scer 12 °P L (t-test p-value > 0.05). There was a significant difference between Scer 12 °P, Scer 16 °P and Scer 20 °P (ANOVA p-value < 0.05), but with no significant difference between Scer 12 °P and Scer 16 °P (t-test p-value > 0.05). The results were similar for the anti-oxidative potential, with no significant difference between Scer 16 °P and Scer 20 °P. The biological activity of the yeast extracts of different yeast strains varied considerably, with (Scer 12 °P) having the highest values, followed by Tdel 12 °P and Spas 12 °P. Significant differences (ANOVA p-value < 0.05) could be established for all the investigated yeast strains in relation to the reductive and anti-oxidative potential. Only the reduction potential of Spas 12 °P and Sfib 12 °P were not significantly different (t-test p-value > 0.05). A wide range of different components such as peptides, vitamins, phenolic components and enzymes were responsible for the biological activity [6,15,49].

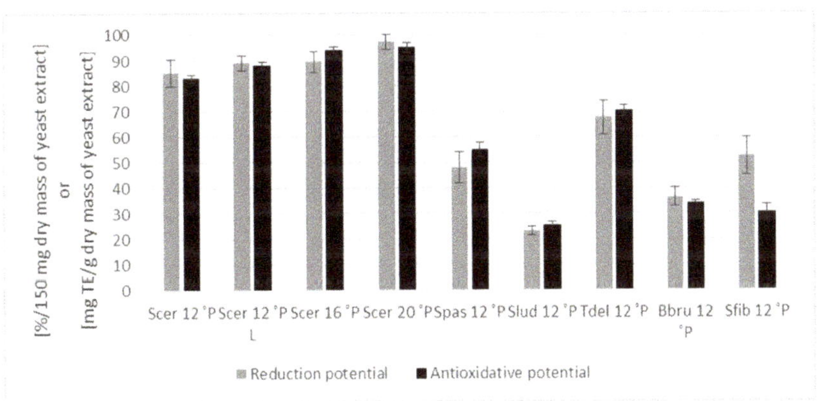

Figure 8. Reduction potential of yeast extract made from spent yeast (*S. cerevisiae* TUM 68) of beer production via mechanical disruption method (sonotrode); influence of original gravity (12 °P, 16 °P, 20 °P), time of yeast cropping (after primary fermentation vs. after cold beer lagering (L)) and yeast strain (Scer, Spas, Slud, Tdel, Bbru, Sfib); for results of ANOVA and pairwise t-test, see text.

4. Conclusions

This study showed that the biodiversity of spent yeast from the brewing process means that it can substantially influence the composition of physiologically important ingredients in the resulting yeast extract. The yeast strain (commercial *Saccharomyces* and alternative non-*Saccharomyces* yeast strains), the original wort content of the fermentation medium and the spent yeast cropping time, have a direct impact on different components of the yeast extract. The general nutrient composition (protein, fat and ash content) of the yeast extracts displayed significant differences after using various spent yeasts. We have evidenced in detail that the release of proteinogenic amino acids during autolysis of the spent yeasts differed greatly and thereby influenced the yeast extracts' FAN. The respective proteinogenic amino acid spectrum also varied. Contamination of the spent yeast with the beer spoiler *L. brevis* had no impact on the amino acid profile of the yeast extracts. Using relevant autolytic process conditions made it possible to increase the GABA concentration in the yeast extract of *S. cerevisiae* TUM 68, however a commercial use in this context is doubtful. It was possible to influence both the total folate content and the proportion of individual folate vitamers by using different yeast strains. Lagering cellar yeast as a starting material to produce yeast extract resulted in lower total folate contents in the yeast extract than primary fermentation yeast. The original wort content of the fermentation medium had no significant influence on the total folate of the spent yeast or the yeast extract. The biological activity (reduction and antioxidative potential) of the yeast extracts also depended on which spent yeast was used from the brewing process. The top-fermenting yeast strain *S. cerevisiae* TUM 68 gave particularly high values. In conclusion, the results indicate that brewer's spent yeast should be carefully selected to produce a yeast extract with a defined nutritional composition. A further research objective would be to adapt the autolytic or mechanical production process for brewer's spent yeast to influence the content of physiologically important ingredients in a yeast extract.

Supplementary Materials: The following are available online at http://www.mdpi.com/2311-5637/5/2/51/s1.

Author Contributions: F.F.J. designed and performed the experiments, analyzed the data and wrote the paper; L.S. and M.R. contributed reagents/materials/method/analysis tools for folate analysis and supported the practical implementation. M.H. selected/provided yeast and bacteria strains and isolated some of the strains in previous studies. M.R., M.H. and F.-J.M. revised the conception and manuscript, and agreed for submission.

Funding: This research received no external funding.

Acknowledgments: The authors thank Martin Neumeier (Technische Universität München–Lehrstuhl für Analytische Lebensmittelchemie, Freising, Germany) and Steffen Pfeil (Technische Universität München–Forschungszentrum Weihenstephan für Brau- und Lebensmittelqualität, Freising, Germany) for technical assistance.

Conflicts of Interest: The authors declare no conflict of interest.

References

1. Annemüller, G.; Manger, H.J.; Lietz, P. *The Yeast in the Brewery*, 1st ed.; VLB Berlin: Berlin, Germany, 2011.

2. Narziss, L.; Back, W. *Die Bierbrauerei Band 2: Die Technologie der Würzebereitung*; WILEY-VCH Verlag GmbH & Co. KGaA: Weinheim, Germany, 2009.

3. Ferreira, I.M.P.L.V.O.; Pinho, O.; Vieira, E.; Tavarela, J.G. Brewer's *Saccharomyces* yeast biomass: Characteristics and potential applications. *Trends Food Sci. Technol.* **2010**, *21*, 77–84. [CrossRef]

4. Gibson, B.R. 125th anniversary review: Improvement of higher gravity brewery fermentation via wort enrichment and supplementation. *J. Inst. Brew.* **2011**, *117*, 268–284. [CrossRef]

5. Jacob, F.F.; Hutzler, M.; Methner, F.-J. Comparison of various industrially applicable disruption methods to produce yeast extract using spent yeast from top-fermenting beer production: Influence on amino acid and protein content. *Eur. Food Res. Technol.* **2019**, *245*, 95–109. [CrossRef]

6. Jacob, F.F.; Striegel, L.; Rychlik, M.; Hutzler, M.; Methner, F.J. Yeast extract production using spent yeast from beer manufacture: Influence of industrially applicable disruption methods on selected substance groups with biotechnological relevance. *Eur. Food Res. Technol.* **2019**, *245*, 1169–1182. [CrossRef]

7. Podpora, B.; Świderski, F.; Sadowska, A.; Piotrowska, A.; Rakowska, R. Spent brewer's yeast autolysates as a new and valuable component of functional food and dietary supplements. *J. Food Process. Technol.* **2015**, *6*, 1.

8. Le Van, V.M.; Strehaiano, P.; Nguyen, D.L.; Taillandier, P. Microbial protease or yeast extract—Alternative additions for improvement of fermentation performance and quality of beer brewed with a high rice content. *J. Am. Soc. Brew. Chem.* **2001**, *59*, 10–16.

9. Casey, G.P.; Magnus, C.A.; Ingledew, W.M. High gravity brewing—nutrient enhanced production of high-concentrations of ethanol by brewing yeast. *Biotechnol. Lett.* **1983**, *5*, 429–434. [CrossRef]

10. Casey, G.P.; Magnus, C.A.; Ingledew, W.M. High-gravity brewing—Effects of nutrition on yeast composition, fermentative ability, and alcohol production. *Appl. Environ. Microbiol.* **1984**, *48*, 639–646.

11. Van Zandycke, S.M.; Fischborn, T. The impact of yeast nutrients on fermentation performance and beer quality. *Tech. Q. Master Brew. Assoc. Am.* **2008**, *45*, 290–293. [CrossRef]

12. Ingledew, W.M.; Sosulski, F.W.; Magnus, C.A. An assessment of yeast foods and their utility in brewing and enology. *J. Am. Soc. Brew. Chem.* **1986**, *44*, 166–170. [CrossRef]

13. Lei, H.J.; Zheng, L.Y.; Wang, C.X.; Zhao, H.F.; Zhao, M.M. Effects of worts treated with proteases on the assimilation of free amino acids and fermentation performance of lager yeast. *Int. J. Food Microbiol.* **2013**, *161*, 76–83. [CrossRef]

14. Procopio, S.; Krause, D.; Hofmann, T.; Becker, T. Significant amino acids in aroma compound profiling during yeast fermentation analyzed by PLS regression. *LWT Food Sci. Technol.* **2013**, *51*, 423–432. [CrossRef]

15. Vieira, F.; Carvalho, J.; Pinto, E.; Cunha, S.; Almeida, A.; Ferreira, I. Nutritive value, antioxidant activity and phenolic compounds profile of brewer's spent yeast extract. *J. Food Compos. Anal.* **2016**, *52*, 44–51. [CrossRef]

16. Pentieva, K.; McNulty, H.; Reichert, R.; Ward, M.; Strain, J.J.; McKillop, D.J.; McPartlin, J.M.; Connolly, E.; Molloy, A.; Kramer, K.; et al. The short-term bioavailabilities of 6S-5-methyltetrahydrofolate and folic acid are equivalent in men. *J. Nutr.* **2004**, *134*, 580–585. [CrossRef] [PubMed]

17. Jacob, F.F.; Michel, M.; Zarnkow, M.; Hutzler, M.; Methner, F.-J. The complexity of yeast extracts and its consequences on the utility in brewing: A review. *BrewingScience* **2019**, *72*, 50–62. [CrossRef]

18. Piddocke, M.P.; Kreisz, S.; Heldt-Hansen, H.P.; Nielsen, K.F.; Olsson, L. Physiological characterization of brewer's yeast in high-gravity beer fermentations with glucose or maltose syrups as adjuncts. *Appl. Microbiol. Biotechnol.* **2009**, *84*, 453–464. [CrossRef] [PubMed]

19. Callejo, M.J.; García Navas, J.J.; Alba, R.; Escott, C.; Loira, I.; González, M.C.; Morata, A. Wort fermentation and beer conditioning with selected non-*Saccharomyces* yeasts in craft beers. *Eur. Food Res. Technol.* **2018**. [CrossRef]

20. Michel, M.; Meier-Dornberg, T.; Jacob, F.; Methner, F.J.; Wagner, R.S.; Hutzler, M. Review: Pure non-*Saccharomyces* starter cultures for beer fermentation with a focus on secondary metabolites and practical applications. *J. Inst. Brew.* **2016**, *122*, 569–587. [CrossRef]

21. Barrette, J.; Champagne, C.P.; Goulet, J. Development of bacterial contamination during production of yeast extracts. *Appl. Environ. Microbiol.* **1999**, *65*, 3261–3263.

22. Berlowska, J.; Dudkiewicz-Kołodziejska, M.; Pawlikowska, E.; Pielech-Przybylska, K.; Balcerek, M.; Czysowska, A.; Kregiel, D. Utilization of post-fermentation yeasts for yeast extract production by autolysis: The effect of yeast strain and saponin from *Quillaja saponaria*. *J. Inst. Brew.* **2017**, *123*, 396–401. [CrossRef]

23. Meier-Dörnberg, T.; Michel, M.; Wagner, R.S.; Jacob, F.; Hutzler, M. Genetic and phenotypic characterization of different top-fermenting *Saccharomyces cerevisiae* ale yeast isolates. *BrewingScience* **2017**, *70*, 9–25.

24. Masuda, K.; Guo, X.F.; Uryu, N.; Hagiwara, T.; Watabe, S. Isolation of marine yeasts collected from the pacific ocean showing a high production of gamma-aminobutyric acid. *Biosci. Biotechnol. Biochem.* **2008**, *72*, 3265–3272. [CrossRef] [PubMed]

25. Naumann, C.; Bassler, R.; Seibold, R.; Barth, C. *Methodenbuch Band 3, die Chemische Untersuchung von Futtermitteln*; VDLUFA-Verlag: Darmstadt, Germany, 1976.

26. Reed, G.; Nagodawithana, T.W. *Yeast Technology*, 2nd ed.; Van Nostrand Reinhold: New York, NY, USA, 1991.

27. Jacob, F. *MEBAK Brautechnische Analysemethoden Würze Bier Biermischgetränke*; Selbstverlag der MEBAK: Freising/Weihenstephan, Germany, 2012.

28. Striegel, L.; Chebib, S.; Netzel, M.E.; Rychlik, M. Improved stable isotope dilution assay for dietary folates using LC-MS/MS and its application to strawberries. *Front. Chem.* **2018**, *6*, 11. [CrossRef] [PubMed]

29. Caballero-Cordoba, G.M.; Sgarbieri, V.C. Nutritional and toxicological evaluation of yeast (*Saccharomyces cerevisiae*) biomass and a yeast protein concentrate. *J. Sci. Food Agric.* **2000**, *80*, 341–351. [CrossRef]

30. Pratt, P.L.; Bryce, J.H.; Stewart, G.G. The effects of osmotic pressure and ethanol on yeast viability and morphology. *J. Inst. Brew.* **2003**, *109*, 218–228. [CrossRef]

31. Casey, G.P.; Chen, E.C.H.; Ingledew, W.M. High-gravity brewing: Production of high levels of ethanol without excessive concentrations of esters and fusel alcohols. *J. Am. Soc. Brew. Chem.* **1985**, *43*, 179–182. [CrossRef]

32. Powell, C.D.; Quain, D.E.; Smart, K.A. The impact of brewing yeast cell age on fermentation performance, attenuation and flocculation. *FEMS Yeast Res.* **2003**, *3*, 149–157. [CrossRef]

33. Steckley, J.D.; Grieve, D.G.; Macleod, G.K.; Moran, E.T. Brewers-yeast slurry, 1. Composition as affected by length of storage, temperature, and chemical treatment. *J. Dairy Sci.* **1979**, *62*, 941–946. [CrossRef]

34. Vieira, E.; Cunha, S.C.; Ferreira, I.M.P.L.V.O. Characterization of a potential bioactive food ingredient from inner cellular content of brewer's spent yeast. *Waste Biomass Valori.* **2018**, 1–8. [CrossRef]

35. Podpora, B.; Swiderski, F.; Sadowska, A.; Rakowska, R.; Wasiak-Zys, G. Spent brewer's yeast extracts as a new component of functional food. *Czech. J. Food Sci.* **2016**, *34*, 554–563. [CrossRef]

36. Reeds, P.J. Dispensable and indispensable amino acids for humans. *J. Nutr.* **2000**, *130*, 1835–1840. [CrossRef] [PubMed]

37. Hans, M.A.; Heinzle, E.; Wittmann, C. Quantification of intracellular amino acids in batch cultures of *Saccharomyces cerevisiae*. *Appl. Microbiol. Biotechnol.* **2001**, *56*, 776–779. [CrossRef] [PubMed]

38. Cooper, D.J.; Stewart, G.G.; Bryce, J.H. Yeast proteolytic activity during high and low gravity wort fermentations and its effect on head retention. *J. Inst. Brew.* **2000**, *106*, 197–201. [CrossRef]

39. Ormrod, I.H.L.; Lalor, E.F.; Sharpe, F.R. The release of yeast proteolytic-enzymes into beer. *J. Inst. Brew.* **1991**, *97*, 441–443. [CrossRef]

40. Fukal, L.; Kas, J.; Rauch, P. Properties of yeast proteinases. *J. Inst. Brew.* **1986**, *92*, 357–359. [CrossRef]

41. Maddox, I.S.; Hough, J.S. Proteolytic enzymes of *Saccharomyces carlsbergensis*. *Biochem. J.* **1970**, *117*, 843–852. [CrossRef]

42. Lenney, J.F.; Dalbec, J.M. Purification and properties of 2 proteinases from *Saccharomyces cerevisiae*. *Arch. Biochem. Biophys.* **1967**, *120*, 42–48. [CrossRef]

43. Schneiderbanger, J.; Grammer, M.; Jacob, F.; Hutzler, M. Statistical evaluation of beer spoilage bacteria by real-time PCR analyses from 2010 to 2016. *J. Inst. Brew.* **2018**, *124*, 173–181. [CrossRef]

44. Ohtomo, M.; Kimura, K.; Watanabe, S.; Toeda, K. Production of components containing gamma-aminobutyric acid from rice bran by *Lactobacillus brevis* IFO12005. *Seibutsu-Kogaku Kaishi* **2006**, *84*, 479–483.

45. Champagne, C.P.; Barrette, J.; Goulet, J. Interaction between pH, autolysis promoters and bacterial contamination on the production of yeast extracts. *Food Res. Int.* **1999**, *32*, 575–583. [CrossRef]

46. Hjortmo, S.; Patring, J.; Jastrebova, J.; Andlid, T. Inherent biodiversity of folate content and composition in yeasts. *Trends Food Sci. Technol.* **2005**, *16*, 311–316. [CrossRef]

47. Hjortmo, S.; Patring, J.; Andlid, T. Growth rate and medium composition strongly affect folate content in *Saccharomyces cerevisiae*. *Int. J. Food Microbiol.* **2008**, *123*, 93–100. [CrossRef] [PubMed]

48. Kanauchi, O.; Igarashi, K.; Ogata, R.; Mitsuyama, K.; Andoh, A. A yeast extract high in bioactive peptides has a blood-pressure lowering effect in hypertensive model. *Curr. Med. Chem.* **2005**, *12*, 3085–3090. [CrossRef] [PubMed]

49. Vieira, E.F.; Melo, A.; Ferreira, I.M.P.L.V.O. Autolysis of intracellular content of brewer's spent yeast to maximize ACE-inhibitory and antioxidant activities. *LWT Food Sci. Technol.* **2017**, *82*, 255–259. [CrossRef]

 *fermentation*

Article

Deoxynivalenol (DON) Accumulation and Nutrient Recovery in Black Soldier Fly Larvae (*Hermetia illucens*) Fed Wheat Infected with *Fusarium* spp.

Zehra Gulsunoglu [1,2], Smitha Aravind [2], Yuchen Bai [2], Lipu Wang [3], H. Randy Kutcher [3] and Takuji Tanaka [2,*]

[1] Faculty of Chemical and Metallurgical, Food Engineering Department, Istanbul Technical University, Istanbul 34469, Turkey
[2] Department of Food and Bioproduct Sciences, University of Saskatchewan, 51 Campus Dr. Saskatoon, SK S7N5A8, Canada
[3] Department of Plant Sciences, University of Saskatchewan, 51 Campus Dr. Saskatoon, SK S7N5A8, Canada
* Correspondence: takuji.tanaka@usask.ca; Tel.: +1-306-966-1697; Fax: +1-306-966-8898

Received: 22 August 2019; Accepted: 16 September 2019; Published: 19 September 2019

Abstract: Fusarium head blight (FHB) is one of the most significant causes of economic loss in cereal crops, resulting in a loss of $50–300 million for Canadian agriculture. The infected grain (containing *Fusarium*-damaged kernels (FDKs)) is often both lower in quality and kernel weight, and it may be unsuitable for human and animal consumption due to mycotoxin presence. However, it still contains a considerable amount of nutrients. A method to recover the nutrients without the mycotoxins should be beneficial for the agricultural economy. In this study, our objective was to examine recovery methods of the nutrients in relation to mycotoxin accumulation in the insect. The FDKs were fermented with *Aspergillus oryzae* and/or *Lactobacillus plantarum* (solid-state fermentation (SSF)). The SSF kernels were then provided to 50 young, black soldier fly larvae (BSFL) for 12 days. Weight gain, chemical composition, and mycotoxin bioaccumulation of BSFL and spent feed were evaluated. After 12 days of insect culture, the BSFL grew 5–6 times their initial weight. While the overall weights did not significantly vary, the proteins and lipids accumulated more in SSF FDK-fed insects. During the active growth period, the larval biomass contained deoxynivalenol (DON), a mycotoxin, at detectable levels; however, by day 12, when the larvae were in the pre-pupal stage, the amount of DON in the insect biomass was nearly negligible, i.e., BSFL did not accumulate DON. Thus, we conclude that the combination of BSFL and SSF can be employed to recover DON-free nutrients from FHB-infected grain to recover value from unmarketable grain.

Keywords: insect culture; solid-state fermentation; mycotoxins; value-added processing; Fusarium head blight

1. Introduction

Fusarium head blight (FHB) is a fungal disease caused by several *Fusarium* spp. Wheat, barley, oats, corn, and other cereal grains can be affected by FHB, resulting in small lightweight kernels and, thus, loss of yield. *Fusarium* spp. produce various amounts and types of trichothecene mycotoxins, which are highly toxic to humans and livestock [1]. A major mycotoxin produced by *Fusarium* spp. is deoxynivalenol (DON). Toxin production occurs during disease development in the field under favorable weather conditions. Contamination of food and feedstuff with DON causes short- and long-term adverse effects on human health and livestock productivity [2]. In order to limit the mycotoxins in food and feed, regulations specify maximum allowable concentrations, which is 1 mg/kg sample in many countries. According to the regulations, products are monitored, and when mycotoxin concentrations exceed the maximum allowable limits, products are separated from the food chain [3]. The economic loss from FHB accounts for many millions of dollars in Canada alone.

Detoxification methods are expensive, labor-intensive, inefficient, and time-consuming, and there is inadequate capacity for industrial applications. An effective way to prevent FHB in the field is to treat the flowering wheat plants with fungicides, and to develop resistant cultivars to minimize the infection of *Fusarium* spp. Fungicide, however, has limited effects on the infection, and, every year, a large number of grains are damaged by FHB. Considering that possible approaches to prevent the contamination of grain with mycotoxins are limited before harvest, alternate approaches should be considered to utilize inedible FHB-damaged kernels (FDK). In this study, we aimed to investigate if black soldier fly larvae (BSFL) can grow on FDK without any toxin accumulation in larval body.

World population is increasing, and it is predicted to reach 9.6 billion by 2050, i.e., a 2.3 billion increase in the next 30 years. Food production relies on agriculture, but the current practice in agriculture may not be sufficient to supply enough food for this population increase, without damaging Mother Earth or introducing super-high-yield crops that do not result in environmental damage. Utilization of inedible agriculture products can bypass the above concerns and can yield additional edible products from current practice agricultural production [4].

There is a considerable interest in the use of insects to recover inedible organic matter because insects can convert carbohydrates into proteins and lipids using organic wastes [5]. *Hermetia illucens* (black soldier fly) is one of the most important species, along with other insect species like *Tenebrio molitor* (yellow mealworm), *Drosophila melanogaster* (common fruit fly), *Amyelois transitella* (orange worm), *Helicoverpa zea* (corn earworm), and *Trichoplusia ni* (cabbage looper) [3]. While the regulations vary among countries and areas, the nutrient values and ease of utilization of insect nutrients draw huge interests for providing an alternative food source [6]. BSFL are considered a possibly proteinaceous animal feed or human food source because of the high accumulation of fat (29%) and protein (42%) in their body, and they do not transmit pathogenic microbes to humans and animals [7]. BSFL have high feed conversion ratios and an ability to convert various organic wastes into body mass [8,9].

Solid-state fermentation (SSF) shows great possibilities in the development of high-value products. Fungal and bacterial strains can be used in SSF, utilizing their abilities of enzyme production such as cellulase, pectinase, and xylanases. In this study, we used *Aspergillus oryzae* and *Lactobacillus plantarum* as microbial strains based on data obtained from our previous study [4]. The main objective of SSF was increasing the bioavailability of nutrients in FDK in favor of recovering them as BSFL biomass, and changing the nutrient profiles in favor of improved/efficient nutrient recovery in BSFL. Through these microbial modifications, efficiency of nutrient recovery from damaged crops should be enhanced during BSFL digestion.

The BSFL can be utilized to reduce pollution and convert low-value organic resources into a high-quality feed protein. They do not harbor diseases, and their production does not need any special equipment of facilities. BSFL are an extremely resistant species capable of dealing with demanding environmental conditions, such as drought, food shortage, or oxygen deficiency. The BSFL is already used in the waste management of some substrates such as manure, rice straw, food waste, kitchen waste, distillers' grains, rotting plant tissues, fecal sludge, animal offal, and animal manure [8]. We showed that BSFL can convert agricultural wastes into biomass with up to 95% recovery of organic matter using SSF [4,10]. Our previous research [4,10] suggests that FDK could be fed to BSFL to recover nutrients at a high rate with SSF treatment of feedstock. It is, however, unclear if the mycotoxins would be accumulated when BSFL are fed with FDK. We hypothesize that the BSFL are not affected by the mycotoxin content of wheat grain infected with *Fusarium* spp., they will not accumulate mycotoxins in their bodies, and a majority of the nutrients in the FDK will be recovered using SSF treatments. Efficient recovery of nutrients from damaged crops can be used to create a high-value product from low-value grain. In this study, the performance of BSFL in converting SSF-treated FDK into insect biomass and the accumulation of DON were investigated.

2. Material and Methods

2.1. Materials

BSFL were purchased from Worm Lady (McGregor, ON, Canada). All chemicals used in this study were commercially available ACS grade and were purchased from Fisher Scientific (Ottawa, ON) and VWR International (Edmonton, AB). *Aspergillus oryzae* NRRL 32657 (*Ao*) and *Lactobacillus plantarum* NRRL B4496 (*Lp*) were obtained from the ARS Culture Collection (USDA, Peoria, IL, USA).

2.2. Solid-State Fermentation

The initial DON concentration of FDK was 0.63 ± 0.20 µg/g dry matter (dm). The FDKs were soaked in water for 18 h at 21 °C to obtain softer kernel for fermentation. The soaked kernels were shred using a household coffee mill (Cuisinart DBM-8, Woodbridge, ON, Canada). Seed cultures of *Ao* and *Lp* were prepared by inoculating Potato Dextrose (PD) and De Man, Rogosa, and Sharpe (MRS) broths. The seed culture of *Ao* was prepared by inoculating 250 mL of PD broth in an Erlenmeyer flask with two loopfuls of spores, followed by 72-h incubation at 30 °C on a rotary shaker (150 rpm). The seed culture of *Lp* was prepared by inoculating 1 mL of fully grown preculture in 250 mL of MRS broth in an Erlenmeyer flask and incubating for 24 h at 37 °C on a rotary shaker (150 rpm). After incubation, fungal and bacterial seed cultures were collected by centrifugation at 6000 rpm for 10 min at 10 °C (Sorvall, RC28S, Manasquan, NJ, US). The biomass obtained was re-suspended in 1/10 of the original volume of sterile distilled water.

Approximately 58 g (34 g in dry weight) of shredded kernels were weighed into each glass jar and, from the seed cultures, 1.5 mL of *Lp*, 2 mL of *Ao*, and 3.5 mL of a combination of these two strains (*Lp* + *Ao*) were inoculated into the crushed kernels. The moisture content was adjusted to 55% (*w/w*) with sterile water. The control sample was prepared without initial seed culture inoculation. The samples were fermented at 30 °C for four days, and the moisture content was kept constant by adding sterile water and mixing once per day under aseptic conditions.

2.3. BSFL Digestion

Fifty BSFL (second instar) were introduced to each SSF FDK sample in the glass jars that had perforated lids to allow moisture and gas transfer during BSFL digestion. Twelve jars were prepared under the same conditions for each SSF FDK sample. The jars were kept at 30 °C for 12 days. Water was added by weighing the sample jars and mixing each day under aseptic conditions.

2.4. BSFL Separation after Digestion

After interval days (0, 4, 8, and 12 days), the larvae were separated from the residual feed using forceps. The larvae were rinsed with water to remove residual substrate from their surface and dried on paper towel; then, their wet weight was determined. Then, the larvae were frozen at −20 °C for further analysis. The rinsed-off feed residues were placed back in the feed bed to avoid errors in spent feed analyses.

2.5. Larval Weight Gain Determination and Survival Rate of Larvae

For survival analysis, the number of larvae was counted at 0, 4, 8, and 12 days (initially, exactly 50 larvae). To monitor larval growth, the dry weight of each sample was determined. Larval volume was calculated by multiplying the length, width, and thickness of 10 individual larvae before and after digestion.

2.6. Proximate Analysis of BSFL and Spent Feed

Proximate analysis was performed on the BSFL and spent feed before and after larval digestion of the fermented FDK. The dry weight of the samples was measured after drying at 105 °C for 24 h to a constant weight according to AOAC Method #930.15 [11]. Dried samples were used for ash, crude protein, and crude fat analyses. The ash content of the larval biomass and spent feed was determined by the gravimetric method as described in AOAC Method #942.05 [11]. Samples were carbonized using a hot plate in the fume hood and, after carbonization, the crucibles were incinerated in a muffle furnace at 550 °C overnight. Crude protein content was measured by the micro-Kjeldahl method as described in AOAC Method #960.52 [11] with slight modification. Conversion factors of 6.25 and 5.70 were used to calculate total protein of the larval biomass and spent feed, respectively. The crude fat analysis was determined according to the Goldfisch method as described in AACC method #30-20.01 [12]. Samples were weighed onto Whatman filter paper (No.1) at 0.5 g for larval biomass and 1 g for spent feed and placed in the Goldfisch apparatus (Labconco Corporation, Kansas City, MO, US). Petroleum ether was used as the extraction solvent, and the extraction process lasted 6 h. Crude fat was determined as the weight of fats in the extract after removal of the solvent. The carbohydrate contents of BSFL and spent feed were determined by subtracting the lipid, protein, and ash contents from the total weight.

2.7. Mycotoxin Analysis

Larval biomass and spent feed were finely ground and extracted for mycotoxin analysis to determine the concentration of DON accumulated. Larval samples and spent feed were extracted according to the method developed by Dr. L. Wang, at the Cereal and Flax Pathology program at the University of Saskatchewan (personal communication). Finely ground wheat grain (2 g) and a larval sample (0.1 g) were mixed with acetonitrile/water (84:16, *v/v*) with a ratio of 1:4 *w/v* and extracted on a rotary shaker for 2 h at room temperature (250 rpm). The extract was diluted 1:10 with 5 mM ammonium acetate and syringe-filtered. Thirty microliters of this filtrate was injected into the LC–MS/MS.

The LC–MS/MS conditions were developed on a high-performance liquid chromatography system (Agilent 1260 Infinity Quaternary; Agilent Technologies, Mississauga, ON, CA) coupled to an AB Sciex 4000 hybrid triple quadrupole linear ion trap (4000 QTrap) mass spectrometer (Concord, ON, CA) equipped with a Turboionspray^{TM} interface. Applied Biosystems/MDS Sciex Analyst software (Version 1.6.2, AB Sciex, Foster City, CA, US) was used for system control and quantification. The mobile phase consisted of a mixture of solvent A (5 mM ammonium acetate in water) and solvent B (5 mM ammonium acetate in methanol). Samples were stored in the auto sampler at 4 °C and a 30-µL injection volume with a 3-s flush port wash (to minimize carryover) was used to introduce the sample into the column.

Chromatographic separation was obtained at a flow rate of 300 µL/min through an Agilent ZORBAX Eclipse XDB C18 column (4.6 × 100 mm, 1.8 µm) equipped with an Eclipse XDB C18 (4.6 mm, 1.8 µm) guard column maintained at 30 °C in a column heater. Multiple reaction monitoring (MRM) with electrospray was used to monitor DON with the transitions of *m/z* 355.0 to *m/z* 296.1 and *m/z* 355.0 to *m/z* 264.9 as quantifier and qualifier ions, respectively.

The DON standard (purity > 99%) was supplied by Romer Labs Inc. (Tulln, Austria), and the stock solution was prepared at a concentration of 1 µg/mL in acetonitrile and kept at −20 °C. Working stock solutions were made by diluting in 5 mM ammonium acetate to the level of 10-fold final working concentration for each standard and quality control (QC) point. Standard and QC samples were prepared by adding 100 µL of each working stock to 900 µL of blank sample and mixing gently. A standard curve of seven points was constructed by determining the best fit of peak area versus the analyte concentration and running a weighed 1/*x* linear regression analysis.

2.8. Statistical Analysis

Each treatment was analyzed in triplicate, and results are presented as an average with standard deviation. Statistical significance of the results was analyzed by one-way analysis of variance (ANOVA) using MINITAB (MINITAB 18, Minitab Inc., Coventry, UK). Treatment means were declared significantly different from each other using Tukey's test at $p < 0.05$.

3. Results and Discussion

3.1. Growth Rate of BSFL

The amount of feed left after periods of BSFL rearing is given in Figure 1. The BSFL consumed about 65, 45, 61, and 72 mg FDK/larva/day during the first four days for the control, *Lp*, *Ao*, and *Lp* + *Ao* groups, respectively. In the next four days, the BSFL consumed 67, 82, 76, and 84 mg FDK/larva/day for the control, *Lp*, *Ao*, and *Lp* + *Ao* groups, respectively. After eight days of consumption of FDK, BSFL growth rates decreased for all treatments. The BSFL consumed 30, 42, 21, and 10 mg FDK/larva/day during the last four days for the control, *Lp*, *Ao*, and *Lp* + *Ao* groups, respectively.

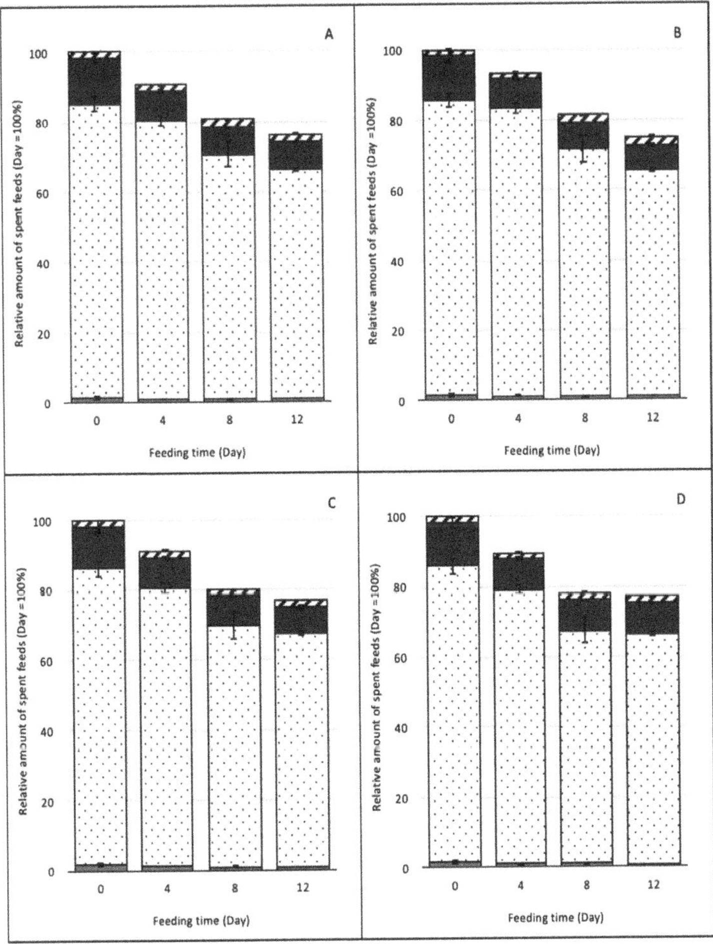

Figure 1. The relative amount of spent feed during black soldier fly larvae (BSFL) digestion. Bars indicate differences in nutrient composition of spent feed after BSFL digestion based on initial day. During feeding time, the residual amount was calculated by taking into account the weight of Fusarium-damaged kernels (FDKs) consumed by BSFL. Each bar was divided according to the ratio of nutrients: hatched area, ashes; black area, proteins; dotted area, carbohydrates; gray area, lipids. Each panel represents the results from (**A**) control without any inoculum (i.e., unfermented FDK), (**B**) fermented FDK with *Lactobacillus plantarum*, (**C**) fermented FDK with *Aspergillus oryzae*, and (**D**) fermented FDK with both *L. plantarum* and *A. oryzae*.

Feed consumption was relative to the growth rate of larvae. The weights and volumes of BSFL from each batch fed FDK fermented with *Lp*, *Ao*, and *Lp* + *Ao* were determined for each feeding time (Figure 2). Increases in weight were observed until day eight for all treatments. From the eighth to the twelveth day, BSFL weight did not differ for any of the treatments. Initial weight of BSFL (day zero) was 8.5 ± 0.4 mg on a dry weight basis (dwb) per larva. Larvae gained significantly more weight when fed with FDK fermented with *Lp* (58.4 ± 6.3 mg/larva). Kuttiyatveetil et al. [4] assessed SSF borage meal and flaxseed meal as the feedstuffs and reported that the highest BSFL biomass was 87.4 mg/larva. The volume of BSFL increased from 66.8 ± 21.8 mm^3 to 352.7 ± 87.9 mm^3 when fed FDK fermented with *Ao* at day eight, and there were no significant differences among treatments. Larval survival rates did not differ among treatments. At the end of the feeding time, 90–93% of initial larvae survived.

It is speculated that the final four days were the pupation period. At prepupa, the last larval stage, the larvae stop feeding in order to produce prothoracicotropic hormone (PPTH), which is necessary for metamorphosis. At this stage, they attain maximum size and have large protein and fat contents to sustain them through metamorphosis; they do not show significant changes in morphological characteristics at this stage [13]. Our results showed low consumption of feeds and weight gains of BSFL, indicating that the last four days can be considered as prepupal periods.

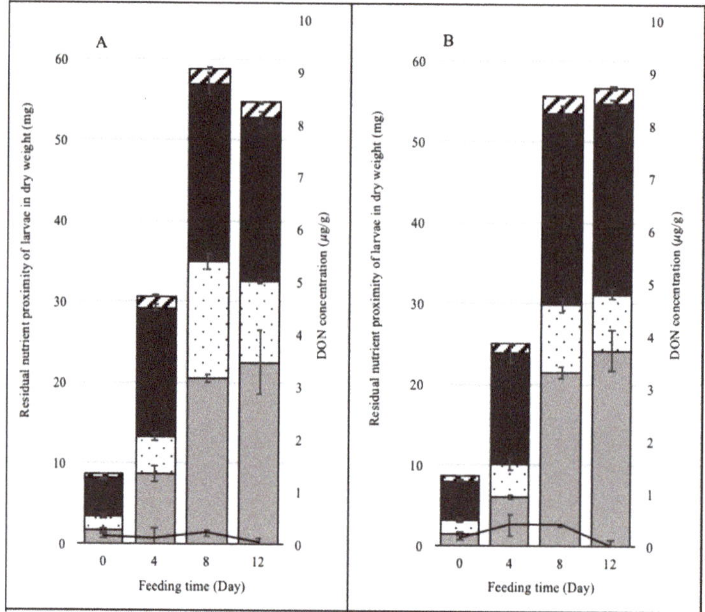

Figure 2. *Cont.*

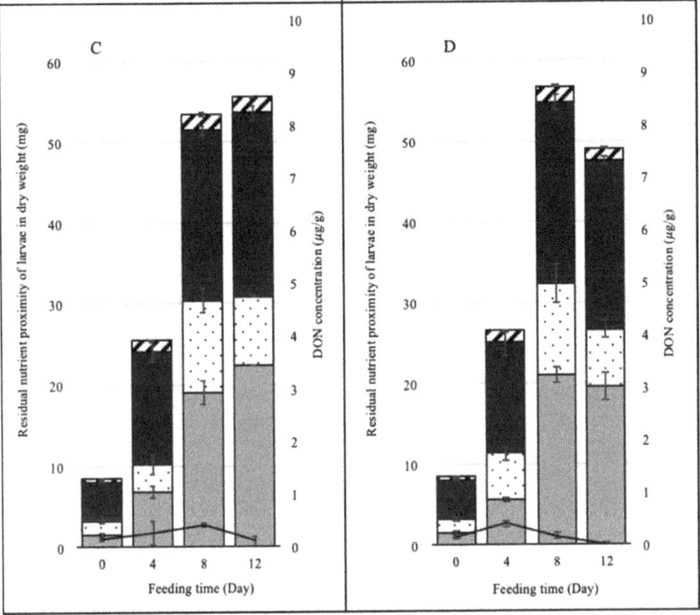

Figure 2. Weight gain, proximate composition, and deoxynivalenol (DON) concentration of larvae during black soldier fly larvae rearing. The overall length of each bar shows the dry weight of larvae divided according to the ratio of nutrients: hatched area, ashes; gray area, lipids; dotted area, carbohydrates; black area, proteins; the lines, DON concentration in black soldier fly larvae. Each panel represents the results of larvae from (**A**) Fusarium-damaged kernels (FDK) without inoculum (i.e., unfermented FDK), (**B**) FDK with *Lactobacillus plantarum*, (**C**) fermented FDK with *Aspergillus oryzae*, and (**D**), fermented FDK with both *L. plantarum* and *A. oryzae*.

3.2. The Proximate Composition of BSFL and Spent Feed

In order to achieve high larval growth, FDK was fermented for four days using generally regarded as safe (GRAS) strains (*Ao* and/or *Lp*). The nutrient profile differences among SSF treatments at day zero indicated that unfermented FDK contained 1.5% lipid, 8.9% protein, 2.2% ash, and 87.7% carbohydrates (Figure 1). At the fourth day, SSF did not change ash and lipid contents; however, the protein and carbohydrate contents differed. The protein content of the grain increased for all fermented FDK treatments; however, the carbohydrate content decreased during SSF for all treatments. While the carbohydrate ratio decreased, its absolute amount remained the greatest in the SSF feed, and the increased protein content was speculated to benefit larval growth. There was no difference in the amount of nutrients among microorganisms tested in this study; however, based on visual observation, FDK became softer compared to control FDK for easier mastication by BSFL.

During BSFL feeding trials, nutrient compositions of spent feed changed (Figure 1). Results showed that BSFL were able to consume and digest carbohydrates (i.e., starch) as their main food sources. The BSFL consumed around 23–25% of the dry matter in SSF feeds in 12 days, i.e., the residual amounts were reduced from 34 g at day zero to 25–26 g at day 12. The protein in spent feed decreased from 4.5 ± 0.5 to 1.8 ± 0.2 g on a dwb of FDK during BSFL digestion after 12 days of digestion for all SSF treatments. Lipid content of spent feed decreased significantly for all treatments after four days of digestion, and the decrease did not change by day 12. The carbohydrate content (i.e., starch) in spent feed decreased significantly during BSFL digestion for all fermented FDK. The ash content of spent feed decreased until the 12th day of digestion for the control, *Ao*, and *Lp* + *Ao* treatments. There was no difference in the ash content of FDK fermented with *Lp*. The decrease in proximate composition of spent feed during BSFL digestion was associated with the consumption of nutrients by BSFL.

The proximate composition of larval biomass was determined (Figure 2). During feeding time, the ash amount of BSFL increased until day eight, when it reached the maximum level for all treatments, and then it remained constant until the 12th day. The initial ash content of BSFL was 0.5 ± 0.0 mg/larva and it reached 2.0 ± 0.0 mg/larva at day eight for FDK fermented with *Lp* + *Ao*. The carbohydrate content of BSFL was 1.7 ± 0.2 mg/larva, and, after the eight-day feeding period, the carbohydrate content increased to 14.6 ± 1.0, 8.3 ± 0.8, 11.3 ± 1.5, and 12.4 ± 1.2 mg/larva fed with FDK fermented in the control, *Lp*, *Ao*, and *Lp* + *Ao* treatments, respectively. The higher carbohydrate contents can be explained by an increase in the volume of BSFL, associated with an increase in larval skin chitin, which is produced during the growing stage of the BSFL. Kaya et al. [14] reported that chitin content increased gradually from larva to adult, and the highest chitin content was observed in adults. The carbohydrate content was higher in the control BSFL than among other treatments ($p < 0.05$). This can be explained by the reduced gain in the protein, lipid, and ash contents, i.e., concentrating of carbohydrates [10]. These results indicated that BSFL can recover the nutrients in damaged FDK and are not affected by the presence of mycotoxins.

Also, it was noticed that the sums of protein and lipid gains were 42.6 ± 4.7, 47.7 ± 3.1, 45.2 ± 0.8, and 40.4 ± 2.1 mg/larva fed with FDK fermented in the control, *Lp*, *Ao*, and *Lp* + *Ao* treatments, respectively, at day 12. At day eight, these figures were 42.3 ± 1.8, 45.2 ± 1.6, 40.1 ± 2.1, and 43.5 ± 1.3, respectively. Meanwhile the accumulation of carbohydrate at day eight (and day 12) were 14.6 ± 1.0 (10.1 ± 0.3), 8.3 ± 0.8 (7.0 ± 0.6), 11.3 ± 1.5 (8.5 ± 0.1), and 11.3 ± 2.3, respectively. These results indicated that SSF grain-fed insects can achieve better protein and lipid profiles compared to unfermented grain-fed insects. It is, therefore, indicated that the SSF assisted to improve the nutrient components of the insect, while overall gain in weight in this study did not significantly vary among the feeds.

3.3. Amounts of DON in Spent Feed and BSFL Biomass

During BSFL digestion, the DON concentration in spent feed at the fourth day was the lowest compared to other days (Table 1). After four days of digestion, the concentration of DON in spent feed continuously increased until day 12. This increase in DON contents of the spent feed was comparable to the 23–25% consumption rate of nutrients by larvae. This suggested that the larvae did not assimilate DON in their body, but it was simply passed through their intestinal system. The highest DON concentration in the spent feed was for FDK fermented with *Lp*.

Table 1. Deoxynivalenol (DON) concentration in spent feed after black soldier fly larvae digestion.

	DON Concentration (µg/g Spent Feed) *			
	Day 0	**Day 4**	**Day 8**	**Day 12**
Control	0.63 ± 0.20 [c,C]	0.15 ± 0.05 [c,A]	4.17 ± 0.83 [b,B]	6.72 ± 0.70 [a,C]
Lactobacillus plantarum	2.60 ± 0.07 [b,B]	0.20 ± 0.07 [c,A]	13.66 ± 0.37 [a,A]	14.17 ± 1.55 [a,A]
Aspergillus oryzae	2.76 ± 0.27 [b,B]	0.12 ± 0.02 [c,A]	5.80 ± 0.88 [a,B]	6.82 ± 1.82 [a,B,C]
L. plantarum + *A. oryzae*	3.58 ± 0.18 [b,A]	0.12 ± 0.05 [b,A]	10.69 ± 1.77 [a,A]	11.41 ± 2.25 [a,A,B]

* Each value is expressed as the mean $\pm$ SD ($n = 3$). Statistical symbols: Means of DON concentration for each fermentation type marked with different lowercase letters (a, b, c) within a row are significantly different between days, and those followed by capital letters (A, B, C) within a column represent significant differences in DON concentration compared with fermentation type on the same day ($p < 0.05$).

There are several possibilities to explain the DON increase. Some microbial enzyme yielded during fermentation could enhance the DON levels, owing to a release of DON from kernel cell walls or other cell components [15]. Secondly, the DON may be produced during the feeding periods. Microorganism type and the proportion of each may result in an increase in mycotoxin production among microorganisms [5]. For example, the increase could also result from the increasing number of *Fusarium* spp. during fermentation time. Goral et al. [16] investigated the relationship between concentration of *Fusarium* biomass and trichothecenes B (DON and nivalenol). They reported a stronger relationship between *Fusarium* biomass and DON content, as confirmed by FHB index. As another possibility, they might be converted from precursors. Nakagawa et al. [17] reported that deoxynivalenol-3-O-glucoside (D3G) has lower toxicity compared to its precursor DON, and D3G can be converted to DON in human or animal gut. Lactic acid bacteria have the capability to hydrolase the derivative D3G back to the toxic DON form. When the lactic acid bacteria ferment the feed, it might be responsible for the chemical conversion of conjugated mycotoxins and explain the high amount of DON in the spent feed [18]. While these factors can affect the results, it is unlikely appropriate to explain our results. We checked the DON amounts at each sampling time for SSF and unfermented grains. Such generations and conversions were counted in the data, regardless of their effect. We did not observe significant differences in DON amounts among the four conditions. Thus, we concluded that the differences in DON amounts between SSF and unfermented FDK were small and they did not significantly affect the contents of DON.

While DON amounts in the spent feed increased during the feeding period, BSFL biomass showed different trends in terms of DON amounts. In the growing stage, the larval biomass contained considerable DON concentration; however, there was no DON accumulation observed in BSFL after the 12-day rearing. Bosch et al. [3] evaluated the tolerance and accumulation of aflatoxin B1 (AFB1) in BSFL fed with AFB1-containing feeds to utilize the mycotoxin-contaminated crops. The BSFL did not contain detectable levels of AFB1 (<0.10 µg/kg). They suggested that the larvae rapidly excreted or metabolized the AFB1 after ingestion. Another possible reason was that part of the AFB1 could be found in bound form with proteins and left undetected. Camenzuli et al. [19] also investigated the potential of accumulation of AFB1, DON, ochratoxin A, and a mixture of mycotoxins in BSFL. None of the mycotoxins accumulated in the larval body; they were shown to excrete or metabolize the four mycotoxins present in the feed. Our results showed that BSFL growing on FDK feeds do not accumulate DON in their body. As shown in Figure 1, the larvae consumed most of the feed during the first eight days, and then consumption stopped by day 12, due to pupation. This is because they defecate their intestinal system and have an empty gut before pupation or shortly after adult emergence. The DON analysis indicated that BSFL do not assimilate DON in their body, and DON observed during the growth stage is speculated to be in the contents of their intestinal system.

4. Conclusions

The FDK used in this study had a large amount of nutrients, and BSFL converted these nutrients to insect biomass without accumulating DON in their bodies. While they consumed DON-contaminated materials, they contained DON in their body; however, the toxins were excreted from their body before they became pupae. They consumed ~2.8 g of FDK per g of BSFL body mass gained, and mainly converted starch into their proteins and lipids at a high efficiency. Thus, the BSFL can be used to separate nutrients from DON in FDK. Proximate analysis of fermented FDK showed higher protein and lipid contents, while there was no significant difference in the BSFL body mass gain among treatments. The procedure can be expanded to other mycotoxin-contaminated materials to recover valuable nutrients wasted in those contaminated materials. This waste treatment technology using BSFL may contribute to reducing the burden of animal protein shortages in the animal feed market and provide new income opportunities for small entrepreneurs in low- and middle-income countries.

Author Contributions: Conceptualization, T.T.; Methodology of fermentation, insect culture and proximate analyses, Z.G., S.A. and T.T.; Methodology of DON analysis, L.W. and H.R.K.; Investigation, Z.G., A.S., Y.B. and L.W.; Writing original draft preparation, Z.G.; Writing-Review & Editing, T.T.; Supervision, T.T.

Funding: This research received no external funding.

Conflicts of Interest: The authors declare no conflict of interest.

References

1. Yoshida, M.; Nakajima, T. Deoxynivalenol and nivalenol accumulation in wheat infected with *Fusarium graminearum* during grain development. *Phytopathology* **2010**, *100*, 763–773. [CrossRef] [PubMed]
2. Yu, Y.H.; Hsiao, F.S.H.; Proskura, W.S.; Dybus, A.; Siao, Y.H.; Cheng, Y.H. An impact of deoxynivalenol produced by *Fusarium graminearum* on broiler chickens. *J. Anim. Physiol. Anim. Nutr.* **2018**, *102*, 1012–1019. [CrossRef] [PubMed]
3. Bosch, G.; Fels-Klerx, H.J.V.; Rijk, T.C.; Oonincx, D. Aflatoxin B1 tolerance and accumulation in black soldier fly larvae (*Hermetia illucens*) and yellow mealworms (*Tenebrio molitor*). *Toxins* **2017**, *9*, 185. [CrossRef] [PubMed]
4. Kuttiyatveetil, J.R.A.; Mitra, P.; Goldin, D.; Nickerson, M.T.; Tanaka, T. Recovery of residual nutrients from agri-food byproducts using a combination of solid-state fermentation and insect rearing. *Int. J. Food Sci. Technol.* **2018**, *54*, 1130–1140. [CrossRef]
5. Schisler, D.A.; Khan, N.I.; Boehm, M.J. Biological control of Fusarium head blight of wheat and deoxynivalenol levels in grain via use of microbial antagonists. *Adv. Exp. Med. Biol.* **2002**, *504*, 53–69. [PubMed]
6. Huis, A.V.; Itterbeeck, J.V.; Klunder, H.; Mertens, E.; Halloran, A.; Muir, G.; Vantomme, P. *Edible Insects: Future Prospects for Food and Feed Security*; FAO Forestry Paper (No. 171); FAO, United Nations: Rome, Italy, 2013.
7. Cickova, H.; Newton, G.L.; Lacy, R.C.; Kozanek, M. The use of fly larvae for organic waste treatment. *J. Waste Manag.* **2015**, *35*, 68–80. [CrossRef] [PubMed]
8. Wang, Y.S.; Shelomi, M. Review of black soldier fly (*Hermetia illucens*) as animal feed and human food. *Foods* **2017**, *6*, 91. [CrossRef] [PubMed]
9. Liu, Z.Y.; Minor, M.; Morel, P.C.H.; Najar-Rodriguez, A.J. Bioconversion of three organic wastes by black soldier fly (Diptera: Stratiomyidae) larvae. *Environ. Entomol.* **2018**, *47*, 1609–1617. [CrossRef] [PubMed]
10. Howdeshell, T.; Tanaka, T. Recovery of glucose from dried distiller's grain with solubles, using combinations of solid-state fermentation and insect culture. *Can. J. Microbiol.* **2018**, *64*, 706–715. [CrossRef] [PubMed]
11. AOAC. *Official Methods of Analysis*, 18th ed.; AOAC International: Rockville, MD, USA, 2005.
12. AACC. *International Approved Methods 1–2*; Cereal and Grain Association: St. Paul, MN, USA, 2009; ISBN 978-1-891127-68-2.
13. Manurung, R.; Suprianta, A.; Rizkita, R.E.; Ramadhani, E.P. Bioconversion of rice straw waste by black soldier fly larvae (*Hermetia illucens* L.): Optimal feed rate for biomass production. *J. Entomol. Zool. Stud.* **2016**, *4*, 1036–1041.
14. Kaya, M.; Sofi, K.; Sargin, I.; Mujtaba, M. Changes in physicochemical properties of chitin at developmental stages (larvae, pupa and adult) of *Vespa crabro* (wasp). *Carbohydr. Polym.* **2016**, *145*, 64–70. [CrossRef] [PubMed]
15. Schaarschmidt, S.; Fauhl-Hassek, C. The fate of mycotoxins during the processing of wheat for human consumption. *Compr. Rev. Food Sci. Food Saf.* **2018**, *17*, 556–593. [CrossRef]
16. Goral, T.; Wisniewska, H.; Ochodzki, P.; Nielsen, L.K.; Walentyn-Goral, D.; Stepien, L. Relationship between Fusarium head blight, kernel damage, concentration of *Fusarium* biomass, and *Fusarium* toxins in grain of winter wheat inoculated with *Fusarium culmorum*. *Toxins* **2019**, *11*, 2. [CrossRef] [PubMed]
17. Nakagawa, H.; He, X.; Matsuo, Y.; Singh, P.K.; Kushiro, M. Analysis of the masked metabolite of deoxynivalenol and *Fusarium* resistance in CIMMYT wheat germplasm. *Toxins* **2017**, *9*, 238. [CrossRef] [PubMed]
18. Berthiller, F.K.R.; Domig, K.J.; Kneifel, W.; Juge, N.; Schuhmacher, R.; Adam, G. Hydrolytic fate of deoxynivalenol-3-glucoside during digestion. *Toxicol. Lett.* **2011**, *206*, 264–267. [CrossRef] [PubMed]
19. Camenzuli, L.; Dam, R.V.; Rijk, T.; Andriessen, R.; Schelt, J.V.; Fles-Klerx, H.J.V. Tolerance and excretion of the mycotoxins aflatoxin B1, zearalenone, deoxynivalenol, and ochratoxin A by *Alphitobius diaperinus* and *Hermetia illucens* from contaminated substrates. *Toxins* **2018**, *10*, 91. [CrossRef] [PubMed]

 *fermentation*

Review

Food Wastes as a Potential New Source for Edible Insect Mass Production for Food and Feed: A review

Vassileios Varelas

Biodynamic Research Institute, Skillebyholm 7, 15391 Järna, Sweden; vavarelas@chem.uoa.gr or vassileios.varelas@sbfi.se

Received: 9 August 2019; Accepted: 27 August 2019; Published: 2 September 2019

Abstract: About one-third of the food produced annually worldwide ends up as waste. A minor part of this waste is used for biofuel and compost production, but most is landfilled, causing environmental damage. Mass production of edible insects for human food and livestock feed seems a sustainable solution to meet demand for animal-based protein, which is expected to increase due to rapid global population growth. The aim of this review was to compile up-to-date information on mass rearing of edible insects for food and feed based on food wastes. The use and the potential role of the fermentation process in edible insect mass production and the potential impact of this rearing process in achieving an environmentally friendly and sustainable food industry was also assessed. Food waste comprises a huge nutrient stock that could be valorized to feed nutritionally flexible edible insects. Artificial diets based on food by-products for black soldier fly, house fly, mealworm, and house cricket mass production have already been tested with promising results. The use of fermentation and fermentation by-products can contribute to this process and future research is proposed towards this direction. Part of the sustainability of the food sector could be based on the valorization of food waste for edible insect mass production. Further research on functional properties of reared edible insects, standardization of edible insects rearing techniques, safety control aspects, and life cycle assessments is needed for an insect-based food industry.

Keywords: edible insects; food wastes; insect mass production; fermentation; sustainability

---

1. Introduction

Entomophagy, i.e., the practice of eating insects as food, formed part of the prehistoric diet in many areas worldwide [1,2]. Over the millennia since then, it has been a regular part of the diet of many people from various cultures throughout the world [3,4]. Globally, more than two billion people, mainly in Asia, Africa, and South America, are estimated to practice entomophagy [2,4,5], with more than 2000 edible insect species being used for this purpose [6]. In Western culture, however, entomophagy is not accepted and is considered a disgusting and primitive behavior, while insects are associated with pests [7]. However, this taboo seems to be weakening in recent years, as eating habits have been changing and a new trend for insect-based products and incorporation of entomophagy into the Western diet has begun [8,9].

In the near future, demand for animal-based food protein is expected to increase by up to 70% [10] due to exponential growth in the global population, which is predicted to reach 9 billion by 2050 [3,8]. The increased food production required to meet this demand will be accompanied by further exhaustion of water, agricultural, forestry, fishery, and biodiversity resources, with negative environmental impacts [11]. When the problem of climate change is added to these concerns, then global food security becomes an even more crucial issue [12,13].

Edible insects are called the insect species which can be used for human consumption but also for livestock feed as a whole, parts of them, and/or protein, and lipid extract [11,14,15]. Edible insects seem a promising alternative solution to achieving food security in the upcoming global food crisis [16], because they provide some significant advantages for human nutrition, including high protein, amino acids, lipids, energy, and various micronutrients [17,18]. Moreover, compared with livestock, insect rearing has a lower environmental impact as multiple and various food sources can be used, greenhouse gas emissions are low, the water and space requirements are low, and the feed conversion rate is high [7,11]. In addition to serving as food and feed, insects can also contribute significantly to food sustainability through biowaste degradation and conversion into food, feed, and fertilizers [19]. Furthermore, they can help preserve biodiversity [20] and assist in plant pollination and pest control [9].

In the global food industry, around 1.3 billion tonnes of various food wastes are discarded every year [21]. The waste generated in the food industry originate mainly from primary production, food processing, wholesale and logistics, combined with retail and markets, food service, and households. For 2012, the estimated volume of food waste for the EU alone was about 88 million tonnes [22]. In the USA, almost 45 million tonnes of fresh vegetables, fruits, milk, and grain products are wasted annually [23]. According to Baiano (2014), up to 42% of total food waste is produced in households [24].

In many cases, food waste residues are difficult to utilize for the recovery of value-added products due to their biological instability, potentially pathogenic nature, high water content, rapid autoxidation, and high level of enzymatic activity [25]. On the other hand, this biomaterial comprises a huge nutrient stock [26] and could be valorized through biodegradation by various edible insect species in a mass production system [9,27,28].

The aim of this review was to compile up-to-date information on rearing edible insects for food and feed purposes using food waste as a substrate. The impact of this bioconversion system in achieving an environmentally friendly and sustainable food industry was also considered.

2. Edible Insect Species Commonly Mass Produced for Food, Feed, and Other Applications

In general, within edible insect rearing and gathering three main strategies are followed: wild harvesting (not farming), semi-domestication (outdoor farming), and farming (indoor farming) [11]. Globally, 92% of edible insect species are wild-harvested, but semi-domestication and farming can provide a food supply in a more sustainable way [3]. Farming of insects for food and feed has recently begun [7].

Regarding consumer acceptance, distribution, rearing conditions, environmental impact, food safety aspects, nutritional value, and use as a component in the diet of farmed animals, pets, and fish, the main commercial edible species harvested in the wild worldwide, but also used for industrial large-scale production, belong to six major orders: Coleoptera, Hymenoptera, Isoptera, Lepidoptera, Orthoptera, and Diptera [15,29].

The most commonly used commercial insects in mass production are mulberry silkworm, waxworm, yellow mealworm, house cricket, black soldier fly, housefly (indoor farming), palm weevil, bamboo caterpillar, weaver ant, grasshopper (outdoor farming), eri silkworm, muga silkworm, giant hornet, and termite (wild farming) [15,30]. The insects most commonly used as animal feed are black soldier fly, housefly, mealworm, beetles, locusts, grasshoppers, crickets, and silkworm [31]. Some edible insect species are also used for medical applications, e.g., *Lucilia sericata* (common green bottlefly) is used as a biological indicator of post-mortem interval (PMI), in human pathology, while its larvae are used in human medicine for healing chronic injuries that cannot be cured with conventional treatments [32]. Moreover, the allantoin secreted by the larvae is used in the treatment of osteomyelitis [30]. Other applications of edible insects include biodegradation of polystyrene in the environment using *Tenebrio molitor* mealworm [33,34], use of black soldier for municipal organic waste management [35], and the use of non-mammalian models like *Galleria mellonella* larvae, also known as waxworm, to model human diseases caused by a number of bacterial pathogens [36].

The most common commercially reared edible insects and their applications for human food and animal and fish feed, as medicines, for component extraction and as environmental treatments are listed in Table 1.

Table 1. Summary of the edible insect species most commonly reared for food and feed, the developmental stage at which they are used, the type of farming system, and commercial applications.

Insect species	Common name	Developmental Stage	Source	Application	Reference
Bombyx mori	Mulberry silkworm	Larvae, pupae	Farming	Human food, animal feed	[11,15,37]
Tenebrio molitor	Yellow mealworm	Larvae	Farming	Human food, feed for pets, zoo animals and fish, polystyrene degradation	[15,28,31,33,34]
Galleria mellonela	Waxworm	Larvae	Farming	Human food, model for human diseases study	[7,30,36]
Rhynchophorus ferrugineus	Red palm weevil	Larvae, pupae	Semi-cultivation	Human food	[38,39]
Rhynchophorus phoenicis	Palm weevil	Larvae	Semi-cultivation	Human food	[39,40]
Acheta domesticus	House cricket	Adult	Farming	Human food, pet food, protein extraction	[15,41]
Gryllus bimacalatus	Mediterranean field cricket	Adult	Farming	Animal feed	[30]
Imbrasia belina	Mopane worm (MW)	Larvae (caterpillar)	Farming	Human food	[42]
Musca domestica	Housefly	Larvae	Farming	Animal and fish feed	[43,44]
Lucilia sericata	Green bottlefly	Larvae(maggot)	Semi-cultivation	Animal and fish feed, Medical treatment	[30–32]
Omphisa fuscidentalis	Bamboo caterpillar	Larvae	Semi-cultivation	Human food	[1,4]
Oecophylla smaragdina	Weaver ant	Adult, larvae, pupae, eggs	Semi-cultivation	Human food, medicine use	[38,45]
Patanga succincta	Grasshopper	Adult	Wild harvesting	Human food	[46]
Oxya spp.	Grasshopper	Adult	Wild harvesting	Human food	[29]
Locusta migratoria	Locust	Adult, nymphs	Farming, wild harvesting	Human food, pet food and fish bait	[11,47,48]
Apis mellifera	Honeybee	Adult	Farming, semi-cultivation	Human food, medical uses (honeybee venom, propolis, royal jelly)	[7,30,49]
Hermetia illucens	Black soldier fly (BSF)	Larvae	Farming	Human food, animal feed	[50]
Macrotermes spp.	Termite	Adult	Wild harvesting	Human food	[51,52]
Encosternum spp.	Stinkbug	Adult	Wild harvesting	Human food	[29,53]
Vespula spp.	(Social) wasp	Larvae	Wild harvesting	Human food	[54,55]
Panchoda marginata	Sun beetle	Larvae	Farming	Human food, animal and fish feed	[7,56]

3. Edible Insect Species That Can Utilize Food Waste as Feed and Their Nutritional Requirements in Mass Production

To date, around 1 million insect species have been described and classified, but the actual number of insect species on Earth is estimated to be between 4 and 30 million. Jongema (2015) compiled a detailed catalogue listing 2037 edible insect species [6], but the actual number of insect species suitable for human food or animal feed applications is still unknown [3].

In recent years, low cost and effective diets, so called artificial diets, are used in lab and/or industry scale in order to rear insects for various purposes (e.g., edible insects, insects as pest predators for pest biological control etc.) [57–59]. Various artificial diets have been introduced for insect rearing, but even the most promising of these is still inferior to natural nutrient sources [60]. The insect species most widely farmed for food and feed purposes are mainly omnivores, which are able to utilize various food sources and thus show broad nutritional flexibility. For this reason, their nutritional requirements and feed rate when fed an artificial diet are difficult to determine [30,61,62]. Due to their nutritional flexibility, the use of low-value food sources can be ideal for large-scale farming of edible insects [11].

A balanced diet composed of organic by-products can be as suitable for the successful growth of mealworm species as the diets used by commercial breeders [28]. It has been reported that an organic food-based diet is critical for larval growth, mass density, and colony maintenance [63]. Recycling of low-quality, plant-derived waste and its conversion into a high-quality feed rich in energy, protein, and fat can be achieved with mealworms in a relatively short time [31]. Moreover, the omnivorous house cricket *Acheta domesticus* can be fed on a large range of organic materials, making it easy to farm in a system producing six or seven generations per year [31].

Most studies with encouraging results regarding artificial diets based on food wastes or mixtures of wastes have been carried out using edible mealworm (*Tenebrio molitor* L., Coleoptera: Tenebrionidae), black soldier fly (*Hermetia illucens*, Diptera: Stratiomyidae), housefly (*Musca domestica*, Diptera: Muscidae), and Cambodian cricket (*Teleogryllus testaceus*, Orthoptera: Gryllidae) and have used raw food material as the insect feed [28,31,60,62,64–67].

Farmed edible insects that utilize food materials and wastes during rearing are summarized in Table 2.

Table 2. Summary of various edible insect species reared on food wastes and their characteristics.

Order	Family	Species	Common Name	Developmental Stage	Degraded Material	Reference
Coleoptera	Tenebrionidae	*Tenebrio molitor* L.	Mealworm	Larvae	Spent grains and beer yeast, bread remains, biscuit remains, potato steam peelings, maize distillers' dried grains with solubles	[28]
Coleoptera	Tenebrionidae	*Tenebrio molitor* L.	Mealworm	Larvae	Mushroom spent corn stover, highly denatured soybean meal, spirit distillers' grains	[64]
Coleoptera	Tenebrionidae	*Zophobas atratus* Fab.	Mealworm	Larvae	Spent grains and beer yeast, bread remains, biscuit remains, potato steam peelings, maize distillers' dried grains with solubles	[28]
Coleoptera	Tenebrionidae	*Alphitobius diaperinus*	Mealworm:	Larvae	Spent grains and beer yeast, bread remains, biscuit remains, potato steam peelings, maize distillers' dried grains with solubles	[28]
Diptera	Stratiomyidae	*Hermetia illucens*	Black soldier fly	Larvae	Waste plant tissues, garden waste, compost tea, catering waste, food scraps	[68]
Diptera	Muscidae	*Musca domestica*	Housefly	Larvae	Mixture of egg content, hatchery waste, and wheat bran	[31]
Orthoptera	Gryllidae	*Acheta domesticus*	House cricket	Adult	Grocery store food waste after aerobic enzymatic digestion, municipal food waste heterogeneous substrate	[41]
Orthoptera	Gryllidae	*Teleogryllus testaceus*	Cambodian field cricket	Adult	Rice bran, cassava plant tops, water spinach, spent grain, residues from mungbean sprout production	[67]

In general, the major macronutrients required for insect mass production are (a) carbohydrates, which serve as an energy pool but are also required for configuration of chitin (exoskeleton of arthropods) [60], (b) lipids (mainly polyunsaturated fatty acids such as linoleic and linolenic), which are the main structural components of the cell membrane, and also store and supply metabolic energy during periods of sustained demands and help conserve water in the arthropod cuticle [29,59,69], and (c) the amino acids leucine, isoleucine, valine, threonine, lysine, arginine, methionine, histidine, phenylalanine, and tryptophan, which insects cannot synthesize [70], and tyrosine, proline, serine, cysteine, glycine, aspartic acid, and glutamic acid, which insects can synthesize, but in insufficient quantities at high energy consumption [61,70]. The essential micronutrients in insect rearing are (a) sterols, which insects cannot synthesize, (b) vitamins, and (c) minerals [30].

The nutrient requirements of edible insects in mass production are summarized in Table 3.

Table 3. Summary of the nutrient requirements of edible insects (adapted from [30,60]).

Macronutrients			Micronutrients		Minerals
Carbohydrates	**Lipids**	**Proteins**	**Sterols *****	**Vitamins**	**Elements ********
Glucose * Fructose * Galactose * Arabinose ** Ribose ** Xylose ** Galactose ** Maltose * Sucrose *	Linoleic (Pfa) *** Linolenic (Pfa) *** Phospholipids ****	Globulins Nucleoproteins Lipoproteins Insoluble proteins Amino acids: Leucine *** Isoleucine *** Valine *** Threonine *** Lysine *** Arginine *** Methionine *** Histidine *** Phenylalanine *** Tryptophan *** Tyrosine **** (major component of sclerotin) Proline **** (important during flight initiation) Serine **** Cysteine **** Glycine **** Aspartic acid **** Glutamic acid ****	Cholesterol Phytosterols (β-sitosterol, campesterol, stigmasterol) Ergosterol	A: Retinol + α-and β-carotene (Ls) B1: Thiamin (Ws) B2: Riboflavin (Ws) B3: Nicotinamide (Ws) B4: Choline (Ws) B5: Pantothenic acid (Ws) B6: Pyridoxine (Ws) B12: Cobalamine (Ws) C: Ascorbic acid (Ls) D: Cholecalsiferol and Ergocalsiferol (Ls) E: α-tocopherol (Ls) K: Phyloquinone (Ls)	Hydrogen Oxygen Carbon Nitrogen Calcium + Phosphorus +++ Chlorine Potassium +++ Sulphur Sodium +++ Magnesium +++ Iron ++ Copper +++ Zinc +++ Silicone Iodine Cobalt Manganese +++ Molybdenum Fluorine Tin Chromium Selenium Vanadium

*: Insects able to absorb and metabolize; **: Insects able to absorb but not metabolize; Pfa: Polyunsaturated fatty acids; ***: Insects unable to synthesize; ****: Insects able to synthesize; Ws: Water-soluble; Ls: Lipid-soluble; *****: Listed in order of importance as essential for living matter (from top down). Minerals consist of combinations of cations and anions of elements; +++: Important for insect growth; ++: Important in enzyme pathways including DNA synthesis; +: Important to a lesser extent, important role in muscular excitation.

Food industry organic wastes are produced in vast quantities and can be valorized for various purposes, e.g., as biofuels, crop fertilizers, pharmaceuticals, functional foods, etc. [25]. The largest quantities are generated by the fruit, vegetable, olive oil, fermentation, dairy, meat, and seafood industries [23]. Food waste comprises a mixture of various food residues, e.g., bread, pastry, noodles, rice, potatoes, meat, and vegetables [21].

Insects are much more efficient at converting feed to body weight than conventional livestock and can be reared on organic waste streams, transforming these into high-value food and feed [31]. The use of food wastes in rearing edible insects is a quite new and promising approach [7,11]. For this purpose, various artificial food waste-based diets covering the nutritional needs of farmed insects have been proposed, without pre-treatment of the biomaterial [28,31,67] (see also Table 2).

The chemical composition and nutritional value of various wastes that have already used in insect rearing are summarized in Table 4.

Table 4. Summary of chemical composition of various food materials and wastes which can be used for rearing edible insects (mainly adapted from DTU-Food database [71]).

Food *	Chemical Composition	Reference
Wheat bran	Total N (2.560%), protein (16.2%), available carbohydrates (24.6%), dietary fiber (40.2%), total fat (5.3%), ash (5.4%), water (8.4%), vitamins (C, E, K1, B1, B2, B3, B5, B6, B9), minerals and inorganics (Na, K, Ca, Mg, P, Fe, Cu, Zn, In, Mn, Cr, Se, Mo, Co, Ni, Cd, Pb), carbohydrates (fructose, glucose, sucrose), saturated fatty acids (C16:0, C18:0, C20:0), monounsaturated fatty acids (C16:1 n-7, C18:1 n-9, C20:1 n-11), polyunsaturated fatty acids (C18:2 n-6, C18:3 n-3, C20:4 n-6), amino acids (isoleucine, leucine, lysine, methionine, cysteine, phenylalanine, tyrosine, threonine, tryptophan, valine, arginine, histidine, alanine, aspartic acid, glutamic acid, glycine, proline, serine)	[71,72]

Table 4. *Cont.*

Food *	Chemical Composition	Reference
Soy flour	Total N (6.520%), protein (37.2%), available carbohydrates (20.2%), dietary fiber (10.4%), total fat (22.2%), ash (5.1%), water (5.1%), vitamins (A, β-carotene, E, K1, B1, B2, B3, B5, B6, B9), minerals and inorganics (Na, K, Ca, Mg, P, Fe, Cu, Z, In, Mn, Cr, Se, Ni, Hg, Cd, Pb), carbohydrates (sucrose, starch, exoses, pentoses, uronic acids, cellulose, lignin), saturated fatty acids (C12:0, C14:0, C16:0, C18:0, C20:0, C22:0), monounsaturated fatty acids (C16:1 n-7, C18:1 n-9, C20:1 n-11), polyunsaturated fatty acids (C18:2 n-6, C18:3 n-3), amino acids (isoleucine, leucine, lysine, methionine, cysteine, phenylalanine, tyrosine, threonine, tryptophan, valine, arginine, histidine, alanine, aspartic acid, glutamic acid, glycine, proline, serine)	[71,73]
Spent grain	Total N (1.890%), protein (11.0%), available carbohydrates (64.3%), dietary fiber (8.5%), total fat (4.2%), water (8.7%), vitamins (B1, B2, B3, B6, B9, E), minerals and inorganics (Na, K, Ca, Mg, P, Fe, Cu, Zn, In, Mn, Cr, Se, Mo, Co, Ni, Hg, Cd, Pb), amino acids (isoleucine, leucine, lysine, methionine, phenylalanine, threonine, tryptophan, valine, arginine, histidine, alanine, aspartic acid, glutamic acid, glycine, proline, serine)	([71,74]
Spent brewer's yeast	Total N (1.340%,%), protein (8.4%), available carbohydrates (12.7%), dietary fiber (6.2%), total fat (1.9%), ash (1.8%), water (69.0%), vitamins (B1, E, B2, B3, B5, B6, B7, B9, C), minerals and inorganics (Na, K, Ca, Mg, P, Fe, Cu, Z, In, Mn, Se, Ni, Cd), carbohydrates (mannose, β-(1,3), (1,6)-glucan, α-(1,4)-glucan, chitin) saturated fatty acids (C12:0, C16:0, C18:0), monounsaturated fatty acids (C16:1 n-7, C18:1 n-9), polyunsaturated fatty acids (C18:2 n-6), amino acids (isoleucine, leucine, lysine, methionine, cysteine, phenylalanine, tyrosine, threonine, tryptophan, valine, arginine, histidine, alanine, aspartic acid, glutamic acid, glycine, proline, serine), nucleic acids	[71,75–77]
Bread remains	Total N (1.400%), protein (8.0%), available carbohydrates (48.0%), dietary fiber (4.0%), total fat (4.3%), ash (1.8%), water (33.9%), vitamins (E, B1, B2, B3, B5, B6, B7, B9), minerals and inorganics (Na, K, Ca, Mg, P, Fe, Cu, Z, In, Mn, Cr, Se, Ni, Hg, As, Cd, Pb), carbohydrates (fructose, glucose, sucrose), saturated fatty acids (C14:0, C16:0, C18:0, C20:0), monounsaturated fatty acids (C16:1 n-7, C18:1 n-9, C20:1 n-11), polyunsaturated fatty acids (C18:2 n-6, C18:3 n-3), amino acids (isoleucine, leucine, lysine, methionine, cysteine, phenylalanine, tyrosine, threonine, tryptophan, valine, arginine, histidine, alanine, aspartic acid, glutamic acid, glycine, proline, serine)	[71,78]
Potato steam peelings	Starch (25%), non-starch polysaccharide (30%), acid insoluble and acid soluble lignin (20%), protein (18%), lipids (1%), and ash (6%),	[71,79]
Potato	Total N (0.324), protein (2.0%), available carbohydrates (15.9%), total fat (0.3%), dietary fiber (1.4%), ash (0.9%), water (79.5%), vitamins (A, B1, B2, B3, B5, B6, B7, B9, C), minerals and inorganics (Na, K, Ca, Mg, P, Fe, Cu, Zn, In, Mn, Cr, Se, Ni, Hg, As, Cd, Pb), carbohydrates (fructose, glucose, sucrose, starch, exoses, pentoses, uronic acids, cellulose), saturated fatty acids (C16:0, C18:0), monounsaturated fatty acids (C16:1 n-7, C18:1 n-9), polyunsaturated fatty acids (C18:2 n-6, C18:3 n-3), amino acids (isoleucine, leucine, lysine, methionine, cysteine, phenylalanine, tyrosine, threonine, tryptophan, valine, arginine, histidine, alanine, aspartic acid, glutamic acid, glycine, proline, serine)	[80]
Dry egg whites	Total N (13.200%), protein (82.3%), available carbohydrates (6.8%), dietary fiber (0.0%), total fat (0.0%), ash (5.1%), water (5.8%), vitamins (B1, B2, B3, B5, B6, B7, B9, B12, D, E), minerals and inorganics (Cl, Na, K, Ca, Mg, P, Fe, Cu, Zn, In, Mn, Cr, Se), amino acids (isoleucine, leucine, lysine, methionine, cysteine, phenylalanine, tyrosine, threonine, tryptophan, valine, arginine, histidine, alanine, aspartic acid, glutamic acid, glycine, proline, serine), cholesterol (16 mg/100 g)	[71,81]
Rice bran	Total N (2.24%), protein (13.4%), available carbohydrates (28.7%), dietary fiber (21.0%), total fat (0.0%), ash (10.0%), water (6.1%), vitamins (B1, B2, B3), minerals and inorganics (Na, K, Ca, P, Fe), carbohydrates (crude fiber 11.5%), amino acids (isoleucine, leucine, lysine, methionine, cysteine, phenylalanine, tyrosine, threonine, tryptophan, valine, arginine, histidine, alanine, aspartic acid, glutamic acid, glycine, proline, serine)	[71,82]

Table 4. *Cont.*

Food *	Chemical Composition	Reference
Carrot	Total N (0.11%), protein (0.7%), available carbohydrates (5.8%), dietary fiber (2.9%), total fat (0.4%), ash (0.7%), water (89.5%), vitamins (A, β-carotene, E, K1, B1, B2, B3, B5, B6, B7, B9, C), minerals and inorganics (Na, K, Ca, Mg, P, Fe, Cu, Zn, In, Mn, Cr, Se, Ni, Hg, As, Cd, Pb), carbohydrates (fructose, glucose, sucrose, hexoses, pentoses, uronic acids, cellulose, lignin), saturated fatty acids (C16:0, C18:0), monounsaturated fatty acids (C18:1 n-9), polyunsaturated fatty acids (C18:2 n-6, C18:3 n-3, C20:4 n-6), amino acids (isoleucine, leucine, lysine, methionine, cysteine, phenylalanine, tyrosine, threonine, tryptophan, valine, arginine, histidine, alanine, aspartic acid, glutamic acid, glycine, proline, serine)	[71,83]
Lettuce	Total N (0.204%), protein (1.3%), available carbohydrates (0.8%), dietary fiber (1.3%), total fat (0.4%), ash (0.8%), water (95.5%), vitamins (A, β-carotene, E, K1, B1, B2, B3, B5, B6, B7, B9, C), minerals and inorganics (Na, K, Ca, Mg, P, Fe, Cu, Zn, In, Mn), carbohydrates (fructose, glucose, sucrose, starch, hexoses, pentoses, uronic acids, cellulose, lignin), saturated fatty acids (C12:0, C16:0, C18:0), monounsaturated fatty acids (C16:1 n-7, C18:1 n-9, C20:1 n-11, C22:1 n-9), polyunsaturated fatty acids (C18:2 n-6, C18:3 n-3, C18:4 n-3, C20:4 n-6, C20:5 n-3, C22:5 n-3, C22:6 n-3), amino acids (isoleucine, leucine, lysine, methionine, cysteine, phenylalanine, tyrosine, threonine, tryptophan, valine, arginine, histidine, alanine, aspartic acid, glutamic acid, glycine, proline, serine)	[71,84]
Cassava plant	Total N (0.218%), protein (1.4%), available carbohydrates (36.3%), dietary fiber (1.8%), total fat (0.3%), ash (0.6%), water (59.7%), vitamins (A, β-carotene, B1, B2, B3, B5, B6, B7, B9, C), minerals and inorganics (Na, K, Ca, Mg, P, Fe, Cu, Zn, In, Mn, Cr, Se, Ni, Hg, As, Cd, Pb), carbohydrates (fructose, glucose, sucrose, starch, hexoses, pentoses, uronic acids, cellulose, lignin), saturated fatty acids (C16:0, C18:0), monounsaturated fatty acids (C16:1 n-7, C18:1 n-9), polyunsaturated fatty acids (C18:2 n-6, C18:3 n-3, C20:4 n-6), amino acids (isoleucine, leucine, lysine, methionine, cysteine, phenylalanine, tyrosine, threonine, tryptophan, valine, arginine, histidine, alanine, aspartic acid, glutamic acid, glycine, proline, serine)	[71,85]
Peanut oil	Total N (0.000%), protein (0.0%), available carbohydrates (27.5%), dietary fiber (0.0%), total fat (72.5%), ash (0.0%), water (0.0%), vitamins (E, γ-tocopherol), minerals and inorganics (Na, K, Ca, Mg, P, Fe, Cu, Zn), saturated fatty acids (C16:0, C18:0, C20:0, C22:0, C24:0), monounsaturated fatty acids (C16:1 n-7, C18:1 n-9, C20:1 n-11), polyunsaturated fatty acids (C18:2 n-6, C18:3 n-3, C22:6 n-3, other fatty acids)	[71,86]

* Data refer to natural products that have not been processed or pre-treated.

4. Rearing Conditions and Insect Mass Technologies

Wild harvesting can potentially lead to depletion of natural insect species [3]. For a sustainable insect farming industry, cost-effective rearing, harvesting, and processing technologies are required [19]. The information required for industrial-scale mass production of insects from biowaste and agricultural organic residues for food and feed purposes is not complete, but much research is being conducted in this field and recent data seem very promising [30] (see also Table 2). The need for lower cost, more environmental friendly, and sustainable nutrient resources for insect mass technologies will increase as the production level increases [30]. In this regard, food biomass waste can comprise a potential source of ingredients for artificial diets used in edible insect industrial production [7,11,54].

The artificial diets used in insect mass production vary from liquid to solid, depending mainly on (a) the nutritional needs of the insect in question in terms of macronutrients, micronutrients and minerals (see also Table 3); (b) the feeding adaptation of the insect, meaning the way that food is processed by the mouthparts before ingestion, as these are adapted to match the feeding needs. Insect species possessing sucking mouthparts are liquid feeders, those possessing biting mouthparts are solid feeders, and those that possess modified sucking mouthparts, so called piercing-sucking insects, are able to pierce the host and suck liquefied animal and/or plant tissues [30,60]; (c) the pre-manufacturing of the artificial diet. Liquid diets can be used after encapsulation using different materials (paraffin, PVC, polyethylene, polypropylene) to mimic artificial eggs, a treatment step needed for their containment and presentation [60], while liquids and slurries can be dried and concentrated so that can be dissolved in water or mixed with other ingredients. Semi-liquids are used in pellet or extruded form which can be ingested by insects with biting mouthparts and also by insects with sucking mouthparts [30]. Solids are presented as a feed mash with grinding and mixing of all raw materials, after pelleting of various raw materials or by extrusion. Solids can also be encapsulated with complex coacervation technology using proteins and polysaccharides [87].

The development of low-cost commercial diets is crucial for edible insect production at industrial scale [19]. In mass production, the mechanical equipment needed in an integrated production process, automation, mechanization, and monitoring technologies for rearing, harvesting, processing, packaging, and delivering edible insects must also be applied, in order to reduce costs and produce safe food products in large-scale quantities [5,19].

5. Nutritional Composition, Ingredient Characterization, and Food Functional Properties of Edible Insect Species

Insect farming conditions, insect developmental stage, the artificial diet selected, and the preparation and processing methods used (e.g., frying, boiling, drying) are factors that affect the nutritional composition of the reared insects [11]. Different diets composed of various food wastes have been reported to result in differences in the nutritional value of mealworm larvae [88]. However, most previous studies provide no details about the artificial diets and conditions used for insect rearing or about the preparation and process stages [29,53,54].

To date, data required in INFOODS/EuroFIR recommendations concerning the nutritional value of most common edible insect ingredients are lacking [29]. These data refer to protein, crude proteins, crude lipids, available carbohydrates, moisture, dry matter, energy, vitamins, and minerals.

The nutrient content of some of the most commonly reared edible insects reared on food wastes, in terms of crude proteins, crude lipids, available carbohydrates, vitamins, and minerals, is summarized in Table 5.

Table 5. Nutritional value of the most common edible insects reared on food materials and wastes.

Insect Species	Common Name	Develop-mental Stage	Crude Protein (% Dry Weight)	Lipids (% Dry Weight)	Carbohydrates, Vitamins, Minerals etc.	General Comments	Reference
Tenebrio molitor	Yellow mealworm	Larvae	70–76%	6–12%	c.a. 10%	Leucine, lysine, methionine + cysteine, threonine, and valine were the limiting amino acids comparing with FAO/WHO requirements. Major fatty acids were linoleic acid (C18:2, 30–38%), oleic acid (C18:1, 24–34%), and palmitic acid (C16:0, 14–17%).	[66]
Tenebrio molitor	Yellow mealworm	Larvae	46.9–48.6%	18.9–27.6%	-	Mealworm species can be grown successfully on diets composed of organic by-products. Diet affects mealworm growth, development, and feed conversion efficiency. Diets high in yeast-derived protein appear favorable with respect to reduced larval development time, reduced mortality, and increased weight gain.	[28]
**Zophobas atratus* Fab.	Mealworm	Larvae	34.2–42.5%	32.8–42.5%	-		[28]
**Alphitobius diaperinus*	Mealworm	Larvae	64.3–65.0%	13.4–21.8%	-		[28]
**Acheta domesticus*	House cricket	Adult	10.2–28.6%	2.2–12.0%	Carbohydrates (as crude fiber): 13.2–28.9% Minerals: - Vitamins: -	It is possible, using very simple means, to rear local field crickets at ambient temperature in Cambodia. Agricultural and food industry by-products tested here also have potential for use as cricket feed, alone or in combination.	[67]

Table 5. *Cont.*

Insect Species	Common Name	Develop-mental Stage	Crude Protein (% Dry Weight)	Lipids (% Dry Weight)	Carbohydrates, Vitamins, Minerals etc.	General Comments	Reference
Acheta domesticus	House cricket	Adult	16%	-	-	Crickets fed the solid filtrate from food waste processed at an industrial scale via enzymatic digestion were able to reach a harvestable size and achieve feed and protein efficiencies. Crickets reared on waste substrates of sufficient quality might be the most promising path for producing crickets economically	[41]
Acheta domesticus	House cricket	Adult	15.6% ± 8.1%	4.56% ± 2.15%	Carbohydrates: - Minerals: Na, Fe, Zn, Ca, I Vitamins: B12, B2	Data show considerable variation within insect species	[29]

The research field concerning characterization of food functional properties of the most common edible insects (e.g., amino acid and lipid composition, foam ability and foam stability, water absorption capacity (WAC), fat absorption capacity (FAC), protein solubility, microstructure and color, rheological properties, etc.) is quite new. Some data is available, mainly for yellow mealworm, silkworm, house cricket, and housefly [54,89–91].

The food functional properties characterized for the most commonly reared edible insects are summarized in Table 6.

Table 6. Ingredient characterization and food functional properties of most common edible insect species.

Insect Species	Common Name	Developmental Stage	Characterization of Food Properties	Reference
Bombyx mori	Silkworm	Pupae	Amino acid analysis, lipid determination	[89]
Tenebrio molitor	Yellow mealworm	Larvae	Amino acid composition (ion exchange chromatography), protein quality (color, protein content, and molecular weight), molecular weight distribution of the insect protein fractions (SDS-PAGE), foam ability and foam stability, rheological properties	[90]
Tenebrio molitor	Yellow mealworm	Larvae	Amino acid composition, water absorption capacity (WAC), fat absorption capacity (FAC), protein solubility, microstructure and color, rheological properties	[91]
Acheta domesticus	House cricket	Adult	Amino acid composition (ion exchange chromatography), protein quality (color, protein content, and molecular weight), molecular weight distribution of the insect protein fractions (SDS-PAGE), foam ability and foam stability, rheological properties	[90]
Musca domestica	Housefly	Pupae	Moisture, protein, fat, ash, acid detergent fiber (ADF), neutral detergent fiber (NDF), minerals, amino acids, fatty acids, vitamins, and selected carotenoid determination	[92]
Apis mellifera	Honeybee	Eggs, larvae, adult	Determination of water content, crude fiber (structural carbohydrates), fat, free nitrogen extract and mineral salts, crude proteins, Vitamin B2	[93]
Hermetia illucens	Black soldier fly	Larvae	Moisture, protein, fat, ash, acid detergent fiber (ADF), neutral detergent fiber (NDF), minerals, amino acids, fatty acids, vitamins, and selected carotenoid determination	[92]

6. Fermentation Process in Edible Insect Chain Production

The fermentation process is applied during the edible insect production to the following stages:

(a) Valorization of food waste via fermentation and then use of edible insects, especially of the black soldier fly (BSF) [94,95]. The use of pre-fermentation can be performed for the waste stabilization and the food safety increasement. Moreover, the pre-fermentation can enhance the digestibility and bioavailability of nutrients to the insect larvae as most nutrients present in agricultural residue or byproducts are found in insoluble form [94]. The solid residues produced by processing of food waste via microaerobic fermentation (MF) and by black soldier fly larvae (BSF) have been proposed as soil fertilizers for plant growth [95].

(b) Use of fermentation by-products and food wastes as ingredients of artificial diets used for edible insect production. The edible mealworm species *Tenebrio molitor* L., *Zophobas atratus* Fab. and *Alphitobius diaperinus Panzer* were grown successfully on diets composed of organic by-products originating from beer brewing, bread/cookie baking, potato processing, and bioethanol production [28]. The *Hermetia illucens* edible insect, commonly named black soldier fly (BSF), was used for the biodegradation of kitchen residues, grass, sewage sludge, and separated solid material from biogas plants [68]. House crickets (*Acheta domesticus*) have been reared on diets based on food waste processed at an industrial scale via enzymatic digestion [41].

(c) Fermentation of the produced edible insect orders to increase the product's shelf-life and minimize the microbial risks for the consumers associated with edible insect consumption [96,97]. Successful acidification and effectiveness in product's safeguarding shelf-life and safety was achieved by the control of *Enterobacteria* and bacterial spores after lactic fermentation of flour/water mixtures with 10% or 20% powdered roasted mealworm larvae [97]. Techniques such as drying, acidifying, and lactic fermentation can preserve edible insects and insect products without the use of a refrigerator [16].

7. Legislation, Food Safety, and Potential Hazards Associated with the Edible Insect Food-to-Food Production Chain

The legislation concerning edible insects for food and feed varies worldwide. Current EU legislation is quite strict, with the application of two regulations: (a) Regulation 2015/2283 (European Food Safety Authority, EFSA) refers to the use of edible insects as food. Since these were not consumed in the EU before March 1997, they were initially considered 'novel foods' [98], while in the reformed regulations they are not specifically mentioned as novel foods [99]. However, if they are intended to be sold on the EU market, they require authorization from the EFSA. (b) Regulation EU 999/2001 refers to the use of edible insects as feed [100]. According to the International Platform on Insects for Food and Feed (IPIFF), only purified insect fat and hydrolyzed insect proteins are allowed to be used as feed for livestock, while non-hydrolyzed insect proteins can currently only be used and sold as pet food and for fur animals feeding while insects derived proteins are not allowed for use in pig or poultry feed [101]. The recent EU regulation No 2017/893 authorizes the use of insect proteins originating from seven insect species: Common Housefly (*Musca domestica*), Black Soldier Fly (*Hermetia illucens*), Yellow Mealworm (*Tenebrio molitor*), Lesser Mealworm (*Alphitobius diaperinus*), House Cricket (*Acheta domesticus*), Banded Cricket (*Gryllodes sigillatus*), and Field Cricket (*Gryllus assimilis*), as feed in aquaculture [99].

Despite the strict regulatory framework, some EU countries are moving rapidly towards approval of edible insects for food and feed purposes [102]. The Netherlands tolerates the sale of edible insects included in the 'List of Edible Insects of the World' [6], while in Belgium the Agence Fédérale pour la Sécurité de la Chaîne Alimentaire (AFSCA) is carrying out a risk analysis on the sale of edible mealworms, crickets, and locusts as novel foods for the Belgian market [102,103]. In Germany, the EU regulation referring to processed animal proteins (PAPs) is interpreted such that insects PAPs are not allowed as feed (not even in aquaculture), as insects are not slaughtered, but this feed ban does not apply to live insects. Therefore a proposal has been made to Deutsche Landwirtschaftsgesellschaft (DLG) to list live insects as a direct animal feed ingredient [103]. In the United Kingdom the law is looser, allowing insects to be sold as food, but this will change relatively soon with a compulsory application procedure required for classification of insect-based food products [103]. In Switzerland, edible insects require authorization from the Federal Office of Food Safety and Veterinary Services (FFSVO) if they are intended to be sold on the open market [103], but recently FFSVO followed Belgium's policy in allowing particular insect species to be sold for food on the Swiss market [102].

In the US, the legislation on edible insects is also strict and more complex. The main authorities are the Food and Drug Administration (FDA), which regulates the industry and coordinates closely with the United States Department of Agriculture (USDA), and the Animal and Plant Health Inspection Service (APHIS) [102]. Concerning food insect-based products, these must conform to the standard practices of all other US foods, including *Salmonella* and *E. coli* testing and, as edible insects are considered food additives, they must follow FDA regulations as described in the Federal Food, Drug, and Cosmetics Act (FFDCA) for Food Additives [104]. All producers of edible insect products must also conform to all FDA manufacturing procedures, known as Good Manufacturing Practice (GMP) ([103].

In Canada, insect-based foods are considered 'novel' and the legislation is complex, as the food safety and public health standards are set by the Canadian Food Inspection Agency (CFIA), which falls under Health Canada, while novel food safety assessments are conducted under the Food Directorate [103]. In Australia and New Zealand, the food safety and hygiene standards are set by Food Standards Australia New Zealand (FSANZ), in which edible insects are classified as 'novel foods' (non-traditional foods), as in EU regulations, and require an assessment of public health and safety issues before their commercialization, unless they are prohibited from sale [102,103].

In Asia, Thailand appears to be a pioneer and one of the most progressive and innovative countries in edible insect mass production, collection, processing, transport, and marketing of cricket (with most farms being medium- or large-scale enterprises) and palm weevil larvae, but also weaver ants, bamboo caterpillars, and grasshoppers, which are collected from the wild or are harvested seasonally [38]. In China, despite its population and economic growth, mass production of edible insects has not yet been established [102].

In Africa, collaborations between African and European companies are being developed on value chain production in rearing, processing, distribution, and consumption of edible insects [103].

Industrial mass production of edible insects for food and feed is associated with the hazards involved in any food production chain, which can mainly be classified into heavy metals, mycotoxins, pesticide residues, and pathogens [16]. During the relevant processes in an insect-based food chain, the associated hazards concerning food safety are of two origins: (a) specific to the species and (b) related to rearing, processing practices, preservation, and/or transport conditions. They are classified into (a) chemical, (b) physical, (c) allergen, and (d) microbial [98,105].

The data concerning the potential hazards associated with a food-to-food production chain based on most common edible insects are summarized in Table 7.

Table 7. Hazards associated with food-to-food edible insect production.

General Hazard	Specific Hazard	Substance	Insect	Problem	Reference
Chemical	Pesticides/fungicides	Organophosphorus pesticides (malathion, sumithion)	Locust	Toxic, carcinogenic	[106]
	Persistent organic pollutants	Polybrominated diphenyl ether (PBDE)	House cricket	Bioaccumulative and toxic	[107]
	Heavy metals	Cd	Mealworm larvae (*Tenebrio molitor*)	Toxic, carcinogenic	[108–110]
		As	*Agrotis infusa* moth (Lepidoptera)	Toxic, carcinogenic	[111]
		Ld	Cricket	Toxic, carcinogenic	[105,112]
		Pb, Zn, Cu, Cd	Insect larvae(not specified)	Toxic, carcinogenic	
	Antibiotics	Chloramphenicol	Silkworm (*Bombyx mori*)	Prohibited use in animal production	[113]
	Insect toxic substances (for defense or repellent purposes, manufactured by the insect itself or accumulated by the insect via its environment or food)	Quinones	Bombardier beetle	-	[105]
		Cyanogenic toxic compounds (linamarin or lotaustralin)	Butterfly	-	[105]
		Melanization process because of the appearance of toxic products	Larvae of *Galleria mellonella* infected by a fungus	-	[105]
		Phenolic compounds: benzoquinone	Tenebrionidae: *Ulomoides dermesetoides*, flour beetles (adults) *Tribolium confusum* and *Tribolium castaneum*	Cytotoxic against the human lung carcinoma epithelial cell line A-549, DNA damage, possible carcinogen	[98]
		Venom (with bristles)	Coleoptera Larvae of *Trogoderma* spp.	Envenomation by dietary route, intestinal trauma due to the bristles found on the insect, ulcerative colitis	[105]
	Antinutritional substances	Hydrocyanic acid	Yam beetle (*Heteroligus meles*)	Anoxia, highly toxic	[114]
		Tannins	Yam beetle (*Heteroligus meles*), ant, termite, cricket, *Zonocerus variegatus* (grasshopper)	Protein precipitation, toxic	[114–116]

Table 7. *Cont.*

General Hazard	Specific Hazard	Substance	Insect	Problem	Reference
Physical	Foreign bodies	Materials from the processes as with any other processed food	-	Choking, injury, toxic, pain, allergy	[105]
	Insect parts	Sting, sharp rostrums, pines, coarse hairs, cuticles, wings	-	Choking, asphyxia, pain, allergy	[102]
Allergen	Insect colorants	Carmine dye	Cochineal insects (*Dactylopius coccus* Costa, *Coccus cacti* L.)	Anaphylaxis, urticarial, erythematous eruption	[98]
	Insect proteins	Lentil pest proteins	Lentil pests (*Bruchus lentis*)	Infestation	[117]
		Cross-reactive proteins: tropomyosin and arginine kinase	Mealworm (*Tenebrio molitor* L.)	Allergic shock	[118]
	Insect enzymes	-	Caterpillars (*Lophocampa caryae*)	Drooling, difficulty swallowing, pain, and shortness of breath	[98]
	Insect allergens	Venom	Bee, wasp, hornet	Anaphylactic shock, pain	[105]
		Chitin	Various edible insect species	Allergic reaction	[105]
Microbial	Parasitics	Human protozoan parasites	Black soldier fly larvae (*Hermetia illucens*)	Intestinal myiases	[119]
		Human protozoan parasites	Cockroaches and some Diptera	Gastrointestinal diseases, toxoplasmoses	[120]
	Bacteria	*Salmonella* *Shigella* *Vibrio* spp. *E. coli* *Yesrinia eneterocolitica* *Campylobacter* *Listeria monocytogenes* *Clostridium perfrigens*	Yellow meal beetle (*Tenebrio molitor*) Desert locust (*Schistocerca gregaria*) Silkmoth (*Bombyx mori*) Cricket (*Acheta domesticus*) Whole locust (*Locusta migratoria*)	Salmonellosis Shigellosis Vibriosis Diarrhea Yesriniosis Campylobacteriosis Listeriosis Clostridial myonecrosis	[121–123]
	Fungi	*Aspergillus, Penicillium, Fusarium, Cladosporium, Phycomycetes*	-	Mycotoxins	[98,105]
	Non-conventional transmissible agents (NCTA)	Prions	*Sarcophaga carnaria* pupae	Scrapie in hamsters	[105]
		Prion proteins	Fly larvae, mites	Scrapie (sheep), mad cow disease (cattle)	[124]

8. Edible Insect Rearing Using Food Wastes: Towards Green and Sustainable Food Waste Management

The organic wastes generated in food industry processes are huge in volume and numerous in type [21,125]. Household food streams also comprise a significant quantity of waste that is not exploited but landfilled, causing environmental damage [126]. In recent years, food waste management has attracted much attention, as these waste products can be valorized with green technologies in a sustainable way [127,128] for the production of renewable chemicals, biomaterials, and biofuels [129].

In recent years, more and more consumers from the USA and various European countries, like Netherlands and Belgium, adopted the entomophagy trend as accepted [130,131].

The utilization of food wastes for edible insect rearing for food and feed seems a promising approach [16] and some of the most common edible insects have already been reared on food wastes with encouraging results (see Table 2). Regarding crickets reared on various food waste streams in a controlled temperature and relative humidity greenhouse, with pre-analyzed ratios of feed substrates (moisture content, total N, crude protein content, acid detergent fiber content, crude fat content, ash content) in order to assess their feed quality, the biomass accumulation was strongly influenced by the quality of the diet [41]. Regarding the rearing of three edible mealworm species (*Tenebrio molitor* L., *Zophobas atratus* Fab., and *Alphitobius diaperinus* Panzer) on food industry organic by-products, the effects of dietary composition on feed conversion efficiency and mealworm crude protein and fatty acid profile were assessed, indicating that larval protein content was not influenced by diet composition while larval fat composition was affected by the used diet to a certain extent [28]. The substitution of diets comprising of mixed grains with agro-food industry by-products can lower the cost of commercial mealworm rearing [64]. During an experimental design for rearing of black soldier flies on various food waste, the weight reduction of rested waste materials was determined, indicating the ability of the black soldier fly to degrade food and plant organic waste [68].

The effect of larval density on food utilization during mealworm *T. molitor* rearing on a determined mixture of food materials was evaluated, thus indicating that although the space considerations in insect mass rearing are important in reducing production costs, crowding larvae to save space may be counterproductive. Additionally, it was demonstrated that increasing larval density impacts negatively on the productivity resulting in a reducing efficiency of food conversion linearly, higher food expenses, and lower biomass production [63].

However, in the most up-to-date experimental trials, the artificial diets, the rearing conditions, the nutritional value of the reared edible insects on food wastes, the yield (in terms of protein, fat content, chitin, etc.), the quality, and also the cost-efficiency of each rearing technique are not determined. Additionally, in none of the referred technologies (see Table 2) is a technical and economical evaluation presented. In addition to this, the up-to-date trials have been applied with simple food mixtures of wastes which in many cases the proportion, chemical composition of the used food materials and wastes and the conditions of the feeding substrate (temperature, humidity, microbial stability, etc.) are not referred, thus resulting in a not standardized insect mass rearing method and technology. That, in the case of the valorization of household food wastes is very critical as they consist of a heterogeneous substrate of various food material [41] and the compilation of a standardized artificial diet based on this appears to be complicated. The compilation of an artificial diet based on simpler food industry mixtures of wastes (e.g., spent grain), seems easier and effective [28]. Finally, clinical trials of reared insects on food materials and wastes have not been performed in humans and animals until now.

9. Conclusions

Edible insects could provide a solution to meeting future increasing demand for animal-based protein. In addition, the sustainability of the food industry sector could be improved through the use of food wastes as new substrates or dietary components in large-scale processes rearing edible insects for human food and animal feed purposes. This bioconversion could also contribute significantly to reducing climate change and the environmental impacts of food and feed production. The first trials on feeding insects with food wastes have produced encouraging results. Prospective candidates for this purpose are the black soldier fly, which has also been tested for municipal organic waste management with very good results, mealworms, houseflies, and house crickets.

Although there are some promising experimental results on the valorization of food wastes for edible insect rearing, further research is needed on the creation of artificial diets based on food by-products for edible insect mass production, isolation, and characterization of the nutrient content of reared insects, techno-economical evaluation of used technology, food-to-food chain safety control evaluation, and life cycle assessments of farmed insect species, in order to enable establishment of a modern insect-based food industry. Additionally, the use of various fermentation by-products (e.g., yeast, bacteria, micro-algae, etc.) as potential materials for rearing edible insects, has been studied a little and not sufficiently, and further research on the combination of fermentation techniques with edible insect rearing technologies is proposed.

Author Contributions: Writing—Original draft, preparation, creation and presentation of the published work, V.V.

Funding: This research received no external funding

Conflicts of Interest: The authors declare no conflict of interest

References

1. Pal, P.; Roy, S. Edible insects: Future of Human food—A Review. *Int. Lett. Nat. Sci.* **2014**, *26*, 1–11. [CrossRef]
2. Van Itterbeeck, J.; van Huis, A. Environmental manipulation for edible insect procurement: A historical perspective. *J. Ethnobiol. Ethnomed.* **2012**, *8*, 1–7. [CrossRef] [PubMed]
3. Yen, A.L. Insects as food and feed in the Asia Pacific region: Current perspectives and future directions. *J. Insects Food Feed.* **2015**, *1*, 33–55. [CrossRef]
4. Yi, C.; He, Q.; Wang, L.; Kuang, R. The Utilization of Insect-resources in Chinese Rural Area. *J. Agric. Sci.* **2010**, *2*, 146–154. [CrossRef]

5.	Rumpold, B.A.; Schlüter, O.K. Potential and challenges of insects as an innovative source for food and feed production. *Innov. Food Sci. Emerg. Technol.* **2013**, *17*, 1–11. [CrossRef]

6.	Jongema, Y. *List of Edible Insects of the World*; Wageningen University and Research, Department of Entomology of Wageningen University: Wageningen, The Netherlands, 2015.

7.	Jansson, A.; Berggren, Å. *Insects as Food—Something for the Future?* Swedish University of Agricultural Sciences (SLU): Uppsala, Sweden, 2015.

8.	Caparros Megido, R.; Gierts, C.; Blecker, C.; Brostaux, Y.; Haubruge, É.; Alabi, T.; Francis, F. Consumer acceptance of insect-based alternative meat products in Western countries. *Food Qual. Pref.* **2016**, *52*, 237–243. [CrossRef]

9.	Rumpold, B.A.; Klocke, M.; Schlüter, O. Insect biodiversity: Underutilized bioresource for sustainable applications in life sciences. *Reg. Environ. Change* **2017**, *17*, 1445–1454. [CrossRef]

10.	Oonincx, D.G.A.B.; de Boer, I.J.M. Environmental Impact of the Production of Mealworms as a Protein Source for Humans—A Life Cycle Assessment. *PLoS ONE* **2012**, *7*, 45–54. [CrossRef]

11.	Van Huis, A.; van Itterbeeck, J.; Klunder, H.; Mertens, E.; Halloran, A.; Muir, G.; Vantomme, P. Edible insects: Future prospects for food and feed security. In *FAO Forestry Paper*; FAO: Rome, Italy, 2013; Volume 171.

12.	Pimentel, D.; Berger, B.; Filiberto, D.; Newton, M.; Wolfe, B.; Karabinakis, E.; Clark, S.; Poon, E.; Abbett, E.; Nandagopal, S. Water Resources: Agricultural and Environmental Issues. *BioScience* **2004**, *54*, 909–918. [CrossRef]

13.	Zewdie, A. Impacts of Climate Change on Food Security: A Literature Review in Sub Saharan Africa. *J. Earth Sci. Clim. Change* **2014**, *5*, 45–54.

14.	Dunkel, F.V.; Paine, C. Chapter 1—Introduction to Edible Insects. In *In Insects as Sustainable Food Ingredients*; Dossey, A.T., Morales-Ramos, J., Guadalupe Roja, M., Eds.; Academic Press: San Diego, CA, USA, 2016; Volume 2, pp. 1–27.

15.	Gahukar, R.T. Chapter 4—Edible Insects Farming: Efficiency and Impact on Family Livelihood, Food Security, and Environment Compared With Livestock and Crops. In *Insects as Sustainable Food Ingredients*; Dossey, A.T., Morales-Ramos, J., Guadalupe Roja, M., Eds.; Academic Press: San Diego, CA, USA, 2016; pp. 85–111.

16.	Van Huis, A. Edible insects contributing to food security? *Agric Food Secur.* **2015**, *4*, 20. [CrossRef]

17.	Defoliart, G.R. Edible insects as minilivestock. *Biodivers Conserv.* **1995**, *4*, 306–321. [CrossRef]

18.	Rumpold, B.A.; Schlüter, O.K. Nutritional composition and safety aspects of edible insects. *Mol. Nutr. Food Res.* **2013a**, *57*, 802–823. [CrossRef] [PubMed]

19.	Kelemu, S.; Niassy, S.; Torto, B.; Fiaboe, K.; Affognon, H.; Tonnang, H.; Maniania, N.K.; Ekesi, S. African edible insects for food and feed: Inventory, diversity, commonalities and contribution to food security. *J. Insects Food Feed.* **2015**, *1*, 103–119. [CrossRef]

20.	DeFoliart, G.R. An overview of the role of edible insects in preserving biodiversity. *Ecol. Food Nutr.* **1997**, *36*, 109–132. [CrossRef]

21.	Pleissner, D.; Lin, C.S.K. Valorisation of food waste in biotechnological processes. *Sustain. Chem. Processes* **2013**, *1*, 13–19. [CrossRef]

22.	Stenmarck, A.; Jensen, C.; Quested, T.; Moates, G. Estimates of European Food Waste Levels. Stockholm, Sweden, 2016. Available online: http://www.eufusions.org/phocadownload/Publications/Estimates%20of%20European%20food%20waste%20levels.pdf (accessed on 3 July 2019).

23.	Kosseva, M.R. Chapter 3: Processing of Food Wastes. In *Advances in Food and Nutrition Research*; Taylor, S.L., Ed.; Academic Press: Heidelberg, Germany, 2009; Volume 58, pp. 57–136.

24.	Baiano, A. Recovery of Biomolecules from Food Wastes—A Review. *Molecules* **2014**, *19*, 14821–14842. [CrossRef]

25.	Russ, W.; Meyer-Pittroff, R. Utilizing Waste Products from the Food Production and Processing Industries. *Crit. Rev. Food Sci. Nutr.* **2004**, *44*, 57–62. [CrossRef] [PubMed]

26.	Duff, S.J.B.; Murray, W.D. Bioconversion of forest products industry waste cellulosics to fuel ethanol: A review. *Bioresour. Technol.* **1996**, *55*, 1–33. [CrossRef]

27.	Lalander, C.H.; Fidjeland, J.; Diener, S.; Eriksson, S.; Vinnerås, B. High waste-to-biomass conversion and efficient *Salmonella spp.* reduction using black soldier fly for waste recycling. *Agron. Sustain. Dev.* **2015**, *35*, 261–271. [CrossRef]

28. van Broekhoven, S.; Oonincx, D.G.A.B.; van Huis, A.; van Loon, J.J.A. Growth performance and feed conversion efficiency of three edible mealworm species (Coleoptera: Tenebrionidae) on diets composed of organic by-products. *J. Insect Physiol.* **2015**, *73*, 1–10. [CrossRef] [PubMed]

29. Payne, C.L.R.; Scarborough, P.; Rayner, M.; Nonaka, K. A systematic review of nutrient composition data available for twelve commercially available edible insects, and comparison with reference values. *Trends Food Sci. Technol.* **2016**, *47*, 69–77. [CrossRef]

30. Cortes Ortiz, J.A.; Ruiz, A.T.; Morales-Ramos, J.A.; Thomas, M.; Rojas, M.G.; Tomberlin, J.K.; Yi, L.; Han, R.; Giroud, L.; Jullien, R.L. Chapter 6—Insect Mass Production Technologies. In *Insects as Sustainable Food Ingredients*; Dossey, A.T., Morales-Ramos, J., Guadalupe Roja, M., Eds.; Academic Press: San Diego, CA, USA, 2016; pp. 154–201.

31. Makkar, H.P.S.; Tran, G.; Heuzé, V.; Ankers, P. State-of-the-art on use of insects as animal feed. *Anim. Feed Sci. Technol.* **2014**, *197*, 1–33. [CrossRef]

32. Rueda, L.C.; Ortega, L.G.; Segura, N.A.; Acero, V.M.; Bello, F. *Lucilia sericata* strain from Colombia: Experimental Colonization, Life Tables and Evaluation of Two Artifcial Diets of the Blowfy Lucilia sericata (Meigen) (Diptera: Calliphoridae), Bogotá, Colombia Strain. *Biol. Res.* **2010**, *43*, 197–203. [CrossRef] [PubMed]

33. Yang, Y.; Yang, J.; Wu, W.-M.; Zhao, J.; Song, Y.; Gao, L.; Yang, R.; Jiang, L. Biodegradation and Mineralization of Polystyrene by Plastic-Eating Mealworms: Part 1. Chemical and Physical Characterization and Isotopic Tests. *Environ. Sci. Technol.* **2015a**, *49*, 12080–12086. [CrossRef] [PubMed]

34. Yang, Y.; Yang, J.; Wu, W.-M.; Zhao, J.; Song, Y.; Gao, L.; Yang, R.; Jiang, L. Biodegradation and Mineralization of Polystyrene by Plastic-Eating Mealworms: Part 2. Role of Gut Microorganisms. *Environ. Sci. Technol.* **2015**, *49*, 12087–12093. [CrossRef] [PubMed]

35. Diener, S.; Studt Solano, N.M.; Roa Gutiérrez, F.; Zurbrügg, C.; Tockner, K. Biological Treatment of Municipal Organic Waste using Black Soldier Fly Larvae. *Waste Biomass Valoriz.* **2011**, *2*, 357–363. [CrossRef]

36. Loh, J.M.S.; Adenwalla, N.; Wiles, S.; Proft, T. *Galleria mellonella* larvae as an infection model for group A streptococcus. *Virulence* **2013**, *4*, 419–428. [CrossRef]

37. Paul, D.; Dey, S. Essential amino acids, lipid profile and fat-soluble vitamins of the edible silkworm *Bombyx mori* (Lepidoptera: Bombycidae). *Int. J. Trop. Insect Sci.* **2014**, *34*, 239–247. [CrossRef]

38. Hanboonsong, Y.; Jamjanya, T.; Durst, P.B. *Six-Legged Livestock: Edible Insect Farming, Collection and Marketing in Thailand*; RAP Publication No. 2013/03; Food and Agriculture Organization (FAO), Regional Office for Asia and the Pacific Bangog: Bangkok, Thailand, 2013.

39. Omotoso, O.T.; Adedire, C.O. Nutrient composition, mineral content and the solubility of the proteins of palm weevil, *Rhynchophorus phoenicis f.* (Coleoptera: Curculionidae). *J. Zhejiang Univ. Sci. B.* **2007**, *8*, 318–322. [CrossRef]

40. Elemo, B.O.; Elemo, G.N.; Makinde, M.; Erukainure, O.L. Chemical Evaluation of African Palm Weevil, *Rhychophorus phoenicis*, Larvae as a Food Source. *J. Insect Sci.* **2011**, *11*, 103–119. [CrossRef] [PubMed]

41. Lundy, M.E.; Parrella, M.P. Crickets Are Not a Free Lunch: Protein Capture from Scalable Organic Side-Streams via High-Density Populations of *Acheta domesticus*. *PLoS ONE* **2015**, *10*. [CrossRef] [PubMed]

42. Hope, R.A.; Frost, P.G.H.; Gardiner, A.; Ghazoul, J. Experimental analysis of adoption of domestic mopane worm farming technology in Zimbabwe. *Dev. S. Afr.* **2009**, *26*, 29–46. [CrossRef]

43. Halloran, A.; Roos, N.; Eilenberg, J.; Cerutti, A.; Bruun, S. Life cycle assessment of edible insects for food protein: A review. *Agron. Sustain. Dev.* **2016**, *36*, 13–19. [CrossRef]

44. Van Zanten, H.H.E.; Mollenhorst, H.; Oonincx, D.G.A.B.; Bikker, P.; Meerburg, B.G.; de Boer, I.J.M. From environmental nuisance to environmental opportunity: Housefly larvae convert waste to livestock feed. *J. Clean. Prod.* **2015**, *102*, 362–369. [CrossRef]

45. Van Itterbeeck, J. Prospects of Semi-Cultivating the Edible Weaver Ant Oecophylla Smaragdina. Ph.D. Thesis, Wageningen University, Wageningen, The Netherlands, 2014.

46. Phiriyangkul, P.; Chutima, S.; Srisomsap, C.; Chokchaichamnankit, D.; Punyarit, P. Effect of Food Thermal Processing on Allergenicity Proteins in Bombay Locust (*Patanga Succincta*). *Int. J. Food Eng.* **2015**, *1*, 23–28. [CrossRef]

47. Bednářová, M.; Borkovcová, M.; Mlček, J.; Rop, O.; Zeman, L. Edible insects—Species suitable for entomophagy under condition of Czech Republic. *Acta Univ. Agric. Silvic. Mendel. Brun.* **2013**, *61*, 587–593. [CrossRef]

48. Oonincx, D.G.A.B.; van der Poel, A.F.B. Effects of diet on the chemical composition of migratory locusts (*Locusta migratoria*). *Zoo Biol.* **2011**, *30*, 9–16. [CrossRef]

49. Chen, P.P.; Wongsiri, S.; Jamyanya, T.; Rinderer, T.E.; Vongsamanode, S.; Matsuka, M.; Sylvester, H.A.; Oldroyd, B.P. Honey Bees and other Edible Insects Used as Human Food in Thailand. *Am. Entomol.* **1998**, *44*, 24–29. [CrossRef]

50. Rumpold, B.A.; Schlüter, O. Nutrient composition of insects and their potential application in food and feed in Europe. *Food Chain* **2014**, *4*, 129–139. [CrossRef]

51. De Figueirêdo, R.E.C.R.; Vasconcellos, A.; Policarpo, I.S.; Romeu, R. Edible and medicinal termites: A global overview. *J. Ethnobiol. Ethnomed.* **2015**, *11: 29*, 13–19.

52. Van Huis, A. Insects as Food in sub-Saharan Africa. *Insect Sci. Its Appl.* **2003**, *23*, 163–185. [CrossRef]

53. Dzerefos, C.M.; Witkowski, E.T.F.; Toms, R. Comparative ethnoentomology of edible stinkbugs in southern Africa and sustainable management considerations. *J. Ethnobiol. Ethnomed.* **2013**, *9*. [CrossRef] [PubMed]

54. Durst, P.B.; Leslie, R.N.; Shono, K. Edible forest insects: Exploring new horizons and traditional practices. In *Forest Insects as Food: Humans Bite Back*; Durst, P.B., Johnson, D.V., Leslie, R.N., Shono, K., Eds.; FAO-Regional Office for Asia and the Pacific: Chiang Mai, Thailand, 2008; pp. 1–4.

55. Mitsuhashi, J. Insects as Traditional Foods in Japan. *Ecol. Food Nutr.* **1997**, *36*, 187–199. [CrossRef]

56. Oonincx, D.G.A.B.; van Itterbeeck, J.; Heetkamp, M.J.W.; van den Brand, H.; van Loon, J.J.A.; van Huis, A. An Exploration on Greenhouse Gas and Ammonia Production by Insect Species Suitable for Animal or Human Consumption. *PLoS ONE* **2011**, *5*, e14445. [CrossRef] [PubMed]

57. Hervet, V.A.D.; Laird, R.A.; Floate, K.D. A Review of the McMorran Diet for Rearing Lepidoptera Species With Addition of a Further 39 Species. *J. Insect Sci.* **2016**, *19*, 1–7. [CrossRef] [PubMed]

58. Sahayraj, K.; Balasubramanian, R. Biological control potential of artificial diet and insect hosts reared *Rhynocoris marginatus* (Fab.) on three pests. *Arch. Phytopathol. Plant. Prot.* **2009**, *42*, 238–247. [CrossRef]

59. Hogsette, J.A. New Diets for Production of House Flies and Stable Flies (Diptera: Muscidae) in the Laboratory. *J. Econ. Entomol.* **1992**, *85*, 2291–2294. [CrossRef]

60. Morales-Ramos, J.A.; Rojas, M.G.; Coudron, T.A. Chapter 7—Artificial Diet Development for Entomophagous Arthropods. In *Mass Production of Beneficial Organisms: Invertebrates and Entomopathogens*; Morales-Ramos, J., Shapiro-Illan, D.I., Guadalupe Roja, M., Eds.; Academic Press: San Diego, CA, USA, 2014; pp. 203–240.

61. Costa-Neto, E.M.; Dunkel, F.V. Chapter 2—Insects as Food: History, Culture, and Modern Use around the World. In *Insects as Sustainable Food Ingredients*; Dossey, A.T., Morales-Ramos, J., Guadalupe Roja, M., Eds.; Academic Press: San Diego, CA, USA, 2016; pp. 29–60.

62. Morales-Ramos, J.A.; Rojas, M.G.; Shapiro-llan, D.I.; Tedders, W.L. Use of Nutrient Self-Selection as a Diet Refining Tool in *Tenebrio molitor* (Coleoptera: Tenebrionidae). *J. Entomol. Sci.* **2013**, *48*, 206–221. [CrossRef]

63. Morales-Ramos, J.A.; Rojas, M.G. Effect of Larval Density on Food Utilization Efficiency of *Tenebrio molitor* (Coleoptera: Tenebrionidae). *J. Econ. Entomol.* **2015**, *108*, 2259–2267. [CrossRef]

64. Zhang, X.; Tang, H.; Chen, G.; Qiao, L.; Li, J.; Liu, B.; Liu, Z.; Li, M.; Liu, X. Growth performance and nutritional profile of mealworms reared on corn stover, soybean meal, and distillers' grains. *Eur. Food Res. Technol.* **2019**, *245*, 1–10. [CrossRef]

65. Aguilar-Miranda, E.D.; López, M.G.; Escamilla-Santana, C.; Barba de la Rosa, A.P. Characteristics of Maize Flour Tortilla Supplemented with Ground *Tenebrio molitor* Larvae. *J. Agric. Food Chem.* **2002**, *50*, 192–195. [CrossRef] [PubMed]

66. Barry, T. Evaluation of the Economic, Social, and Biological Feasibility of Bioconverting Food Wastes With the BlackSoldier Fly *(Hermetia illucens)*. Doctoral thesis, University of North Texas, Denton, TX, USA, 2004.

67. Miech, P.; Berggren, Å.; Lindberg, J.E.; Chhay, T.; Khieu, B.; Jansson, A. Growth and survival of reared Cambodian field crickets (*Teleogryllus testaceus*) fed weeds, agricultural and food industry by-products. *J. Insects Food Feed.* **2016**, *2*, 285–292. [CrossRef]

68. Žáková, M. Hermetia illucens application in management of selected types of organic waste. In Proceedings of the 2nd Electronic International Interdisciplinary Conference, Section 12, 2–6 September 2013. Available online: http://www.eiic.cz/archive/?vid=1&aid=2&kid=20201-33 (accessed on 2 September 2013).

69. Chapman, R.F. *The Insects: Structure and Function*, 4th ed.; Cambridge University Press: New York, NY, USA, 1998.

70. Lundgren, J.G. Nutritional aspects of non-prey foods in the life histories of predaceous Coccinellidae. *Biol. Control.* **2009**, *51*, 294–305. [CrossRef]

71. Fooddata-DTU. Technical University of Denmark, 2015. Available online: https://frida.fooddata.dk/?lang=en (accessed on 3 August 2019).

72. Onipe, O.O.; Jideani, A.I.O.; Beswa, D. Composition and functionality of wheat bran and its application in some cereal food products. *Int. J. Food Sci. Technol.* **2015**, *50*, 2509–2518. [CrossRef]

73. Li, S.; Zhu, D.; Li, K.; Yang, Y.; Lei, Z.; Zhang, Z. Soybean Curd Residue: Composition, Utilization, and Related Limiting Factors. *ISRN Ind. Eng.* **2013**. [CrossRef]

74. Mussatto, S.I. Brewer's spent grain: A valuable feedstock for industrial applications. *J. Sci. Food Agric.* **2014**, *94*, 1264–1275. [CrossRef]

75. Aguilar-Uscanga, B.; François, J.M. A study of the yeast cell wall composition and structure in response to growth conditions and mode of cultivation. *Lett. Appl. Microbiol.* **2003**, *37*, 268–274. [CrossRef]

76. Fesel, P.H.; Zuccaro, A. β-Glucan: Crucial component of the fungal cell wall and elusive MAMP in plants. *Fungal Genet. Biol.* **2016**, *90*, 53–60. [CrossRef]

77. Kwiatkowski, S.; Thielen, U.; Glenney, P.; Moran, C. A Study of *Saccharomyces cerevisiae* Cell Wall Glucans. *J. Inst. Brew.* **2009**, *115*, 151–158. [CrossRef]

78. Mousa, E.I.; Al-Mohizea, I.S.; Al-Kanhal, M.A. Chemical composition and nutritive value of various breads in Saudi Arabia. *Food Chem.* **1992**, *43*, 259–264. [CrossRef]

79. Liang, S.; McDonald, A.G. Chemical and Thermal Characterization of Potato Peel Waste and Its Fermentation Residue as Potential Resources for Biofuel and Bioproducts Production. *J. Agric. Food Chem.* **2014**, *62*, 8421–8429. [CrossRef] [PubMed]

80. Izmirlioglu, G.; Demirci, A. Enhanced Bio-Ethanol Production from Industrial Potato Waste by Statistical Medium Optimization. *Int. J. Mol. Sci.* **2015**, *16*, 24490–24505. [CrossRef] [PubMed]

81. Ma, S.; Zhao, S.; Zhang, Y.; Yu, Y.; Liu, J.; Xu, M. Quality characteristic of spray-drying egg white powders. *Mol. Biol. Rep.* **2013**, *40*, 5677–5683. [CrossRef] [PubMed]

82. Nantiyakul, N.; Furse, S.; Fisk, I.; Foster, T.J.; Tucker, G.; Gray, D.A. Phytochemical Composition of *Oryza sativa* (Rice) Bran Oil Bodies in Crude and Purified Isolates. *J. Am. Oil Chem. Soc.* **2012**, *89*, 1867–1872. [CrossRef]

83. Sharma, K.D.; Karki, S.; Thakur, N.S.; Attri, S. Chemical composition, functional properties and processing of carrot—A review. *J. Food Sci. Technol.* **2012**, *49*, 22–32. [CrossRef]

84. Fahad, S.M.; Islam, A.F.M.M.; Ahmed, M.; Uddin, N.; Alam, M.R.; Alam, M.F.; Khalik, M.F.; Hossain, M.S.; Hossain, M.L.; Abedin, M.J. Determination of Elemental Composition of Malabar spinach, Lettuce, Spinach, Hyacinth Bean, and Cauliflower Vegetables Using Proton Induced X-Ray Emission Technique at Savar Subdistrict in Bangladesh. *BioMed Res. Int.* **2015**. [CrossRef]

85. Aro, S.O.; Aletor, V.A.; Tewe, O.O.; Agbede, J.O. Nutritional potentials of cassava tuber wastes: A case study of a cassava starch processing factory in south-western Nigeria. *Livestock Res. Rural Dev.* **2010**, *22*. Available online: http://www.lrrd.org/lrrd22/11/aro22213.htm (accessed on 5 June 2019).

86. Zhao, X.; Chen, J.; Du, F. Potential use of peanut by-products in food processing: A review. *J. Food Sci. Technol.* **2012**, *49*, 521–529. [CrossRef]

87. Xiao, Z.; Liu, W.; Zhu, G.; Zhou, R.; Niu, Y. A review of the preparation and application of flavour and essential oils microcapsules based on complex coacervation technology. *J. Sci. Food Agric.* **2014**, *94*, 1482–1494. [CrossRef] [PubMed]

88. Ramos-Elorduy, J.; González, E.A.; Hernández, A.R.; Pino, J.M. Use of *Tenebrio molitor* (Coleoptera: Tenebrionidae) to Recycle Organic Wastes and as Feed for Broiler Chickens. *J. Econ. Entomol.* **2002**, *95*, 214–220. [CrossRef] [PubMed]

89. Tomotake, H.; Katagiri, M.; Yamato, M. Silkworm Pupae (*Bombyx mori*) Are New Sources of High Quality Protein and Lipid. *J. Nutr. Sci. Vitaminol.* **2010**, *56*, 446–448. [CrossRef] [PubMed]

90. Yi, L.; Lakemond, C.M.M.; Sagis, L.M.C.; Eisner-Schadler, V.; van Huis, A.; van Boekel, M.A.J.S. Extraction and characterisation of protein fractions from five insect species. *Food Chem.* **2013**, *141*, 3341–3348. [CrossRef] [PubMed]

91. Zhao, X.; Vázquez-Gutiérrez, J.L.; Johansson, D.P.; Landberg, R.; Langton, M. Yellow Mealworm Protein for Food Purposes—Extraction and Functional Properties. *PLoS ONE* **2016**, *11*. [CrossRef] [PubMed]

92. Finke, M.D. Complete Nutrient Content of Four Species of Feeder Insects. *Zoo Biol.* **2013**, *32*, 27–36. [CrossRef]

93. Banjo, A.D.; Lawal, O.A.; Songonuga, E.A. The nutritional value of fourteen species of edible insects in southwestern Nigeria. *Afr. J. Biotechnol.* **2006**, *5*, 298–301.

94. Low, Y.; Wein, L. Reversing the nutrient drain through urban insect farming—Opportunities and challenges. *Bioengineering* **2018**, *5*, 226–237. [CrossRef]

95. Alattar, M.A.; Alatter, F.N.; Popa, R. Effects of microaerobic fermentation and black soldier fly larvae food scrap processing residues on the growth of corn plants (*Zea mays*). *Plant Sci. Today* **2016**, *3*, 57–62. [CrossRef]

96. Borremans, A.; Lenaerts, S.; Crauwels, S.; Lievens, B.; Van Campenhout, L. Marination and fermentation of yellow mealworm larvae (*Tenebrio molitor*). *Food Control.* **2018**, *92*, 47–52. [CrossRef]

97. Klunder, H.C.; Wolkers-Rooijackers, J.; Korpela, J.M.; Nout, M.J.R. Microbiological aspects of processing and storage of edible insects. *Food Control.* **2012**, *26*, 628–631. [CrossRef]

98. Belluco, S.; Losasso, C.; Maggioletti, M.; Alonzi, C.C.; Paoletti, M.G.; Ricci, A. Edible Insects in a Food Safety and Nutritional Perspective: A Critical Review. *Compr. Rev. Food Sci. Food Saf.* **2013**, *12*, 296–313. [CrossRef]

99. European Union (EU). Regulation (EU) 2015/2283 on novel foods, amending Regulation (EU) No 1169/2011 of the European Parliament and of the Council and repealing Regulation (EC) No 258/97 of the European Parliament and of the Council and Commission Regulation (EC) No 1852/2001. *Official Journal of the European Union.* 2015. Available online: http://data.europa.eu/eli/reg/2015/2283/oj (accessed on 1 August 2019).

100. European Union (EU). Regulation (EU) 999/2001 on laying down rules for the prevention, control and eradication of certain transmissible spongiform encephalopathies. *Official Journal of the European Union.* 2001. Available online: https://eur-lex.europa.eu/legal-content/EN/ALL/?uri=CELEX%3A32001R0999 (accessed on 1 August 2019).

101. International Platform of Insects for Food and Feed (IPIFF). Available online: http://ipiff.org/insects-eu-legislation (accessed on 1 August 2019).

102. Marone, P.A. Chapter 7—Food Safety and Regulatory Concerns. In *Insects as Sustainable Food Ingredients*; Dossey, A.T., Morales-Ramos, J., Guadalupe Roja, M., Eds.; Academic Press: San Diego, CA, USA, 2016.

103. Halloran, A. Discussion Paper: Regulatory Frameworks Influencing Insects as Food and Feed. Food and Agriculture Organization (FAO), 2014. Available online: http://www.fao.org/edible-insects/39620-04ee142dbb758d9a521c619f31e28b004.pdf (accessed on 21 June 2019).

104. GPO-U.S. Government Publishing Office. Available online: https://www.gpo.gov/fdsys/pkg/USCODE-2010-title21/pdf/USCODE-2010-title21-chap9-subchapIV-sec348.pdf (accessed on 2 August 2019).

105. ANSES-French Agency for Food, Environmental and Occupational Health & Safety. Opinion on: The Use of Insects as Food and Feed and the Review of Scientific Knowledge on the Health Risks Related to the Consumption of Insects. ANSES Opinion Request No. 2014-SA-0153; 2015. Available online: https://www.anses.fr/en/system/files/BIORISK2014sa0153EN.pdf (accessed on 27 June 2019).

106. Saeed, T.; Dagga, F.A. Analysis of residual pesticides present in edible locusts captured in kuwait. *Arab Gulf J. Sci. Res.* **1993**, *11*, 1–5.

107. Gaylor, M.O.; Harvey, E.; Hale, R.C. House crickets can accumulate polybrominated diphenyl ethers (PBDEs) directly from polyurethane foam common in consumer products. *Chemosphere* **2012**, *86*, 500–505. [CrossRef] [PubMed]

108. Lindqvist, L.; Block, M. Excretion of cadmium during moulting and metamorphosis in *Tenebrio molitor*; (Coleoptera; Tenebrionidae). *Comp. Biochem. Physiol. C: Pharmacol. Toxicol. Endocrinol.* **1995**, *111*, 325–328.

109. Pedersen, S.A.; Kristiansen, E.; Andersen, R.A.; Zachariassen, K.E. Cadmium is deposited in the gut content of larvae of the beetle *Tenebrio molitor* and involves a Cd-binding protein of the low cysteine type. *Comp. Biochem. Physiol. C: Toxicol. Pharmacol.* **2008**, *148*, 217–222.

110. Vijver, M.; Jager, T.; Posthuma, L.; Peijnenburg, W. Metal uptake from soils and soil–sediment mixtures by larvae of *Tenebrio molitor* (L.) (Coleoptera). *Ecotoxicol. Environ. Saf.* **2003**, *54*, 277–289. [CrossRef]

111. Green, K.; Broome, L.; Johnston, S.; Heinze, D. Long distance transport of arsenic by migrating Bogong moths from agricultural lowlands to mountain ecosystems. *Vic. Nat.* **2001**, *118*, 112–116.

112. Zhuang, P.; Zou, H.; Shu, W. Biotransfer of heavy metals along a soil-plant-insect-chicken food chain: Field study. *J. Environ. Sci.* **2009**, *21*, 849–853. [CrossRef]

113. Cappellozza, S.; Saviane, A.; Tettamanti, G.; Squadrin, M.; Vendramin, E.; Paolucci, P.; Franzetti, E.; Squartini, A. Identification of *Enterococcus mundtii* as a pathogenic agent involved in the "flacherie" disease in *Bombyx mori* L. larvae reared on artificial diet. *J. Invertebr. Pathol.* **2011**, *106*, 386–393. [CrossRef]

114. Adesina, A.J. Proximate and anti-nutritional composition of two common edible insects: Yam beetle (*Heteroligus meles*) and palm weevil (*Rhynchophorus phoenicis*). *Elixir Food Sci.* **2012**, *49*, 9782–9786.

115. Adeduntan, S.A. Nutritional and antinutritional characteristics of some insects foraging in Akure forest reserve Ondo State, Nigeria. *J. Food Technol.* **2005**, *3*, 563–567.

116. Omotoso, O.T. Nutrient Composition, Mineral Analysis and Anti-nutrient Factors of *Oryctes rhinoceros* L. (Scarabaeidae: Coleoptera) and Winged Termites, *Marcrotermes nigeriensis* Sjostedt. (Termitidae: Isoptera). *Br. J. Appl. Sci. Technol.* **2015**, *8*, 97–106. [CrossRef]

117. Armentia, A.; Lombardero, M.; Blanco, C.; Fernández, S.; Fernández, A.; Sánchez-Monge, R. Allergic hypersensitivity to the lentil pest Bruchus lentis. *Allergy* **2006**, *61*, 1112–1116. [CrossRef] [PubMed]

118. Verhoeckx, K.C.; van Broekhoven, S.; Gaspari, M.; de Hartog-Jager, S.C.; de Jong, G.; Wichers, H.; van Hoffen, E.; Houben, G.; Knulst, A.C. House dust mite (Derp 10) and crustacean allergic patients may be at risk when consuming food containing mealworm proteins. *Clin. Transl. Allergy* **2013**, *3*, 12–16. [CrossRef]

119. Sehgal, R.; Bhatti, H.P.; Bhasin, D.K.; Sood, A.K.; Nada, R.; Malla, N.; Singh, K. Intestinal myiasis due to *Musca domestica*: A report of two cases. *Jpn. J. Infect. Dis.* **2002**, *55*, 191–193.

120. Graczyk, T.K.; Knight, R.; Tamang, L. Mechanical Transmission of Human Protozoan Parasites by Insects. *Clin. Microbiol. Rev.* **2005**, *18*, 128–132. [CrossRef]

121. Marshall, D.L.; Dickson, J.S.; Nguyen, N.H. Chapter 8—Ensuring Food Safety in Insect Based Foods: Mitigating Microbiological and Other Foodborne Hazards. In *Insects as Sustainable Food Ingredients*; Dossey, A.T., Morales-Ramos, J., Guadalupe Roja, M., Eds.; Academic Press: San Diego, CA, USA, 2016; pp. 223–253.

122. Schabel, H.G. Forest insects as food: A global review. In *Edible Forest Insect: Humans Bite Back. Proceedings of a Workshop on Asia-Pacific Resources and Their Potential for Development*; Durst, P.B., Johnson, D.V., Leslie, R.N., Shono, K., Eds.; FAO-Regional Office for Asia and the Pacific: Bangkok, Thailand, 2010.

123. Fraqueza, M.J.R.; da Silva Coutinho Patarata, L.A. Constraints of HACCP Application on Edible Insect for Food and Feed. 2017. Available online: http://dx.doi.org/10.5772/intechopen.69300 (accessed on 1 July 2019).

124. Lupi, O. Could ectoparasites act as vectors for prion diseases? *Int. J. Dermatol.* **2003**, *42*, 425–429. [CrossRef]

125. Kosseva, M.R. Chapter 3—Sources, Characterization, and Composition of Food Industry Wastes. In *Food Industry Wastes*; Kosseva, M.R., Webb, C., Eds.; Academic Press: San Diego, CA, USA, 2013; Volume 3, pp. 37–60.

126. Arancon, R.A.D.; Lin, C.S.K.; Chan, K.M.; Kwan, T.H.; Luque, R. Advances on waste valorization: New horizons for a more sustainable society. *Energy Sci. Eng.* **2013**, *1*, 53–71. [CrossRef]

127. Luque, R.; Clark, J.H. Valorisation of food residues: Waste to wealth using green chemical technologies. *Sustain. Chem. Processes.* **2013**, *1*, 112–116. [CrossRef]

128. Lin, C.S.K.; Koutinas, A.A.; Stamatelatou, K.; Mubofu, E.B.; Matharu, A.S.; Kopsahelis, N.; Pfaltzgraff, L.A.; Clark, J.H.; Papanikolaou, S.; Kwan, T.H. Current and future trends in food waste valorization for the production of chemicals, materials and fuels: A global perspective. *Biofuels Bioprod. Biorefin.* **2014**, *8*, 686–715. [CrossRef]

129. Woon, K.S.; Lo, I.M.C.; Chiu, S.L.H.; Yan, D.Y.S. Environmental assessment of food waste valorization in producing biogas for various types of energy use based on LCA approach. *Waste Manag.* **2016**, *50*, 290–299. [CrossRef] [PubMed]

130. House, J. Consumer acceptance of insect-based foods in the Netherlands: Academic and commercial implications. *Apetite* **2016**, *107*, 47–58. [CrossRef] [PubMed]

131. Hartmann, C.; Siegrist, M. Insects as food: Perception and acceptance. Findings from current research. *Ernahrungs Umschau* **2016**, *64*, 44–50. [CrossRef]

MDPI
St. Alban-Anlage 66
4052 Basel
Switzerland
Tel. +41 61 683 77 34
Fax +41 61 302 89 18
www.mdpi.com

Fermentation Editorial Office
E-mail: fermentation@mdpi.com
www.mdpi.com/journal/fermentation